RIEMANNIAN GEOMETRY

RIEMANNIAN GEOMETRY
A Beginner's Guide
Second Edition

Frank Morgan

Department of Mathematics
Williams College
Williamstown, Massachusetts

Illustrated by James F. Bredt

A K Peters
Wellesley, Massachusetts

Editorial, Sales, and Customer Service Office

A K Peters, Ltd.
289 Linden Street
Wellesley, MA 02181

Manuscript typed by Dan Robb.

Student editorial committee: Adrian Banner, Ian Budden, Heath Dill, David McMath, and Brian Munson.

Library of Congress Cataloging-in-Publication Data

Morgan, Frank.
    Riemannian Geometry : a beginner's guide / Frank Morgan :
      illustrated by James F. Bredt.
      p.   cm.
    Includes bibliographical references (p. –) and index.
    ISBN 1-56881-073-3
    1. Geometry, Riemannian.   I. Title.
  QA685.M76   1997   97-35094
  516.3'.73—DC21   CIP

Printed in the United States of America
02 01 00 99 98   10 9 8 7 6 5 4 3 2 1

This book is dedicated to my teachers — notably Fred Almgren, Clem Collins, Arthur Mattuck, Mabel Riker, my mom, and my dad. Here as a child I got an early geometry lesson from my dad.

F.M.

Contents

Preface

The complicated formulations of Riemannian geometry present a daunting aspect to the student. This little book focuses on the central concept— curvature. It gives a naive treatment of Riemannian geometry, based on surfaces in $\mathbf{R}^n$ rather than on abstract Riemannian manifolds.

The more sophisticated intrinsic formulas follow naturally. Later chapters treat hyperbolic geometry, general relativity, global geometry, and some current research on energy-minimizing curves and the isoperimetric problem. Proofs, when given at all, emphasize the main ideas and suppress the details that otherwise might overwhelm the student.

This book grew out of graduate courses I taught on tensor analysis at MIT in 1977 and on differential geometry at Stanford in 1987 and Princeton in 1990, and out of my own need to understand curvature better for my work. The last chapter includes research by Williams undergraduates. I want to thank my students, notably Alice Underwood; Paul Siegel, my teaching assistant for tensor analysis; and participants in a seminar at Washington and Lee led by Tim Murdoch.

This second edition includes many more exercises and new sections on the isoperimetric problem (Section 9.7) and double Wulff crystals (Section 10.7). I would like to thank Frank Jones, Rob Kusner, John Sullivan, Jean Taylor, and Dave Witte for helpful comments. Following a useful practice of J. C. C. Nitsche [Nit], the bibliography now includes cross-references to each citation herein.

Other books I have found helpful include Laugwitz's *Differential and Riemannian Geometry* [Lau], Hicks's *Notes on Differential Geometry* [Hic] (unfortunately out of print), Spivak's *Comprehensive Introduction to Differential Geometry* [Spi], and Stoker's *Differential Geometry* [Sto]. An excellent, more sophisticated text is provided by Chavel's recent *Riemannian Geometry: A Modern Introduction* [Cha].

I am currently using this book and *Geometric Measure Theory: A Beginner's Guide* [Mor4], both so happily edited by Klaus Peters and illustrated by Jim Bredt, as texts for an advanced, one-semester undergraduate course at Williams.

Williamstown, Massachusetts   Frank Morgan
Frank.Morgan@williams.edu
http://www.williams.edu/Mathematics/fmorgan

CHAPTER 1
Introduction

The central concept of Riemannian geometry is *curvature*. It describes the most important geometric features of racetracks and of universes. We will begin by defining the curvature of a racetrack. Chapter 7 uses general relativity's interpretation of mass as curvature of space to predict the mysterious precession of Mercury's orbit.

The curvature κ of a racetrack is defined as the rate at which the direction vector $\mathbf{T}$ of motion is turning, measured in radians or degrees per meter. The curvature is big on sharp curves, zero on straightaways. See Figure 1.1.

A two-dimensional surface, such as the surface of Figure 1.2, can curve different amounts in different directions, perhaps upward in some directions, downward in others, and along straight lines in between. The principal curvatures κ_1 and κ_2 are the most upward (positive) and the most downward (negative), respectively. For the saddle of Figure 1.2, it appears that at the origin $\kappa_1 = \frac{1}{4}$ and $\kappa_2 = -1$. The *mean curvature* $H = \kappa_1 + \kappa_2 = -\frac{3}{4}$. The *Gauss curvature* $G = \kappa_1 \kappa_2 = -\frac{1}{4}$. At the south pole of the unit sphere of Figure 1.3, $\kappa_1 = \kappa_2 = 1, H = 2$, and $G = 1$.

Since κ_1 and κ_2 measure the amount that the surface is curving in space, they cannot be measured by a bug confined to the surface. They are "extrinsic" properties. Gauss made the astonishing discovery, however,

1

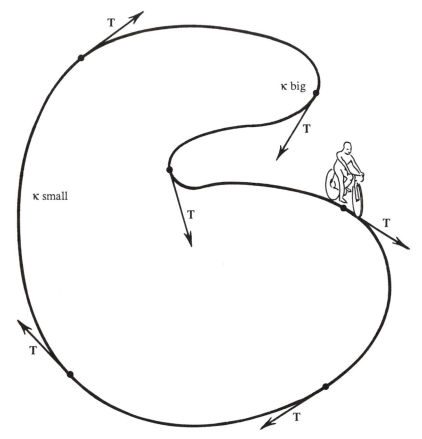

Figure 1.1. Curvature is defined as the rate of change of the direction vector.

that the Gauss curvature $G = \kappa_1\kappa_2$ can, in principle, be measured from within the surface. This result, known as his *Theorema Egregium* or Remarkable Theorem, says that Gauss curvature is an "intrinsic" property. An m-dimensional hypersurface in $\mathbf{R}^{m+1}$ has m principal curvatures $\kappa_1, \ldots, \kappa_m$ at each point. For an m-dimensional surface in $\mathbf{R}^n$, the situation is still more complicated; it is described not by numbers or by vectors, but by the second fundamental *tensor*. Still, Gauss's Theorema Egregium generalizes to show that an associated "Riemannian curvature tensor" is intrinsic.

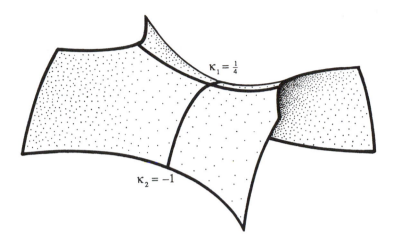

Figure 1.2. At the center of this saddle, the maximum upward curvature is $\kappa_1 = \frac{1}{4}$ and the maximum downward curvature is $\kappa_2 = -1$.

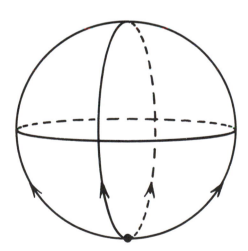

Figure 1.3. At the south pole, the curvature is $+1$ in all directions.

Modern graduate texts in differential geometry strive to give intrinsic curvatures intrinsic definitions, which ignore the ambient $\mathbf{R}^n$ from the outset. In this text, surfaces will start out sitting in $\mathbf{R}^n$, where we can give concrete definitions of the second fundamental tensor and the Riemannian curvature tensor. Only later will we prove that the Riemannian curvature tensor actually is intrinsic.

CHAPTER 2

Curves in $\mathbf{R}^n$

The central idea of Riemannian geometry—curvature—appears already for space curves in this chapter. For a parameterized curve $\mathbf{x}(t)$ in $\mathbf{R}^n$, with velocity $\mathbf{v} = \dot{\mathbf{x}}$ and unit tangent $\mathbf{T} = \mathbf{v}/|\mathbf{v}|$, the curvature vector κ is defined as the rate of change of $\mathbf{T}$ with respect to arc length:

$$\kappa = d\mathbf{T}/ds = \frac{d\mathbf{T}/dt}{ds/dt} = \frac{1}{|\mathbf{v}|}\,\dot{\mathbf{T}}. \tag{2.1}$$

The curvature vector κ points in the direction in which $\mathbf{T}$ is turning, orthogonal to $\mathbf{T}$. Its length, the scalar curvature $\kappa = |\kappa|$, gives the rate of turning. See Figure 2.1. For a planar curve with unit normal $\mathbf{n}$,

$$\kappa = |d\mathbf{n}/ds|. \tag{2.2}$$

For a circle of radius a, κ points toward the center, and $\kappa = 1/a$. For a general curve, the best approximating, or osculating, circle has radius $1/\kappa$, called the radius of curvature.

If the curve is parameterized by arc length, then the curvature vector κ simply equals $d^2\mathbf{x}/ds^2$. If the curve is the graph $\mathbf{y} = f(x)$ of a function $f : \mathbf{R} \to \mathbf{R}^{n-1}$ tangent to the x-axis at the origin 0, then

$$\kappa(0) = f''(0) \in \mathbf{R}^{n-1} \subset \mathbf{R} \times \mathbf{R}^{n-1}.$$

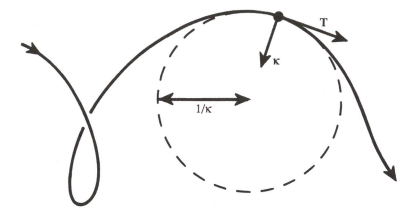

Figure 2.1. The curvature vector κ tells us which way the unit tangent vector $\mathbf{T}$ is turning and how fast. Its length $|\kappa|$ is the reciprocal of the radius of the osculating circle.

Without the tangency hypothesis, the scalar curvature

$$\kappa = |f''|\sqrt{1 + |f'|^2 \sin^2 \theta}/(1 + |f'|^2)^{3/2},$$

where θ is the angle between f' and f''. In $\mathbf{R}^2$, $f : \mathbf{R} \to \mathbf{R}$, $\theta = 0$ of course, and

$$\kappa = |f''|/(1 + |f'|^2)^{3/2}.$$

Curvature tells how the length of a curve changes as the curve is deformed. If an infinitesimal piece of planar curve ds is pushed a distance du in the direction of κ, the length changes by a factor of $1 - \kappa \, du$. Indeed, the original arc lies to second order on a circle of radius $1/\kappa$, and the new one on a circle of radius $1/\kappa - du = (1/\kappa)(1 - \kappa \, du)$. See Figure 2.2. More generally, if the displacement is a vector $d\mathbf{u}$ not necessarily in the direction of κ, only the component of $d\mathbf{u}$ in the κ direction matters, and the length changes by a factor of $1 - \kappa \cdot d\mathbf{u}$. Hence the initial rate of change of length of a curve C in $\mathbf{R}^n$ with initial velocity $\mathbf{V} = d\mathbf{u}/dt$ is $-\int \kappa \cdot \mathbf{V} ds$ (see Section 10.4).

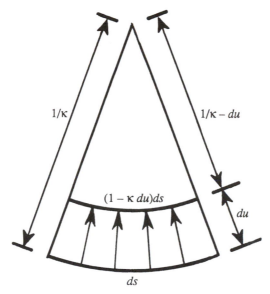

Figure 2.2. An element of arc length ds pushed in the direction of κ decreases by a factor of $1 - \kappa\,du$.

2.1. The smokestack problem. One day I got a call from a company constructing a huge smokestack, which required the attachment of a spiraling strip, or *strake*, for structural support (see Figures 2.3 and 2.4). Of course they had to cut the strake pieces out of a flat piece of metal (see Figure 2.5). The question was, What choice of inner radius r would make the strake fit on the smokestack best?

The curve along which the strake attaches to the smokestack is a helix:

$$\mathbf{x} = (x, y, z) = (a\cos t, a\sin t, ht/2\pi),$$

with the x and y coordinates following a circle of radius $a = 3.75$ feet while the z coordinate increases at a constant rate. In each revolution, θ increases by 2π and z increases by $h = 31.5$ feet. Hence the velocity is

$$\mathbf{v} = \dot{\mathbf{x}} = (-a\sin t, a\cos t, h/2\pi),$$

and the speed is

$$\frac{ds}{dt} = |\mathbf{v}| = \sqrt{a^2 + \frac{h^2}{4\pi^2}} = c \approx 6.26.$$

Figure 2.3. The metal strip, or strake, spirals around the smokestack on a helical path.

The length of one revolution is

$$L = \int_0^{2\pi} |\mathbf{v}| dt = 2\pi c.$$

By analogy with a circle, an engineer guessed that the ideal inner cutting radius r would be $L/2\pi = c \approx 6.26$ feet. When he built a little model, however, he discovered that his guess was too small. After some trial and error, he found that strake pieces cut with $r \approx 10\frac{1}{2}$ feet fit well.

The way to compute the ideal r is to require the strake to have the right *curvature*. We will now compute the curvature κ of the helix and take r to be the radius of curvature $1/\kappa$ (that is, the radius of the circle with the same curvature).

The unit tangent vector $\mathbf{T} = \mathbf{v}/|\mathbf{v}| = \mathbf{v}/c$. Hence the curvature vector is

$$\boldsymbol{\kappa} = d\mathbf{T}/ds = \frac{d\mathbf{T}/dt}{ds/dt} = \frac{\dot{\mathbf{v}}/c}{c} = \frac{1}{c^2}(-a\cos t, -a\sin t, 0),$$

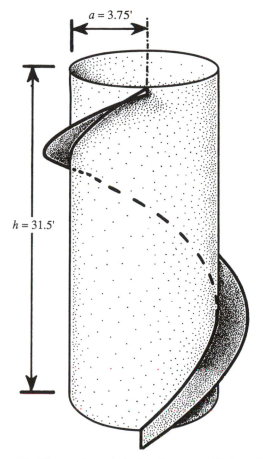

Figure 2.4. The smokestack had radius $a = 3.75$ feet. Each revolution of the strake had a height of 31.5 feet.

and the scalar curvature is $\kappa = a/c^2$. Therefore the ideal inner radius is

$$r = 1/\kappa = c^2/a \approx 10.45 \, \text{feet},$$

in close agreement with the engineer's experiment.

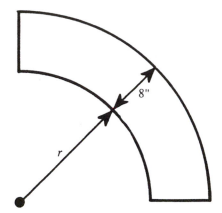

Figure 2.5. When pieces of strake are being cut out of a flat piece of metal, what inner radius r will make the strake fit best along the smokestack?

EXERCISES

2.1. Find the curvature vector at $x = 0$ of

   a. $y = 7x^2 + 8x^3 + 9x^4$

   b. $y = 6x + 7x^2 + 8x^3 + 9x^4$

   c. $y = 5 + 6x + 7x^2 + 8x^3 + 9x^4$

2.2. An "antimoth's" position in $\mathbf{R}^3$ is given by

$$\mathbf{x} = (t\cos t, t\sin t, \frac{2\sqrt{2}}{3}t^{3/2}).$$

   a. Sketch the path.

   b. Compute the curvature vector κ as in the smokestack problem in Section 2.1. (It gets messy.)

   c. Compute the distance traveled in the first four seconds.

CHAPTER 3

Surfaces in $\mathbf{R}^3$

This chapter studies the curvature of a C^2 surface $S \subset \mathbf{R}^3$ at a point $p \in S$. (C^2 just means that locally, the surface is the graph of a function with continuous second derivatives. Computing the curvature will involve differentiating twice.) Let T_pS denote the *tangent space* of vectors tangent to S at p. Let $\mathbf{n}$ denote a unit normal to S at p. *To study the curvature of S, we slice S by planes containing $\mathbf{n}$ and consider the curvature vector $\boldsymbol{\kappa}$ of the resulting curves.* (See Figure 3.1.) Of course each such $\boldsymbol{\kappa}$ must be a multiple of $\mathbf{n}$: $\boldsymbol{\kappa} = \kappa\mathbf{n}$. (For now we will allow κ to be positive or negative. The sign of κ depends on the choice of unit normal $\mathbf{n}$.) It will turn out that the largest and the smallest curvatures κ_1, κ_2 (called the principal curvatures) occur in orthogonal directions and determine the curvatures in all other directions.

Choose orthonormal coordinates on $\mathbf{R}^3$ with the origin at p, S tangent to the x, y-plane at p, and $\mathbf{n}$ pointing along the positive z-axis. Locally S is the graph of a function $z = f(x, y)$. Any unit vector $\mathbf{v}$ tangent to S at p, together with the unit normal vector $\mathbf{n}$, spans a plane, which intersects S in a curve. The curvature κ of this curve, which we call the curvature in the direction $\mathbf{v}$, is just the second derivative

11

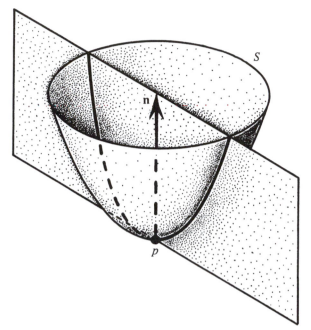

Figure 3.1. The curvature of a surface S at a point p is measured by the curvature of its slices by planes.

$$\kappa = (D^2 f)_p(\mathbf{v}, \mathbf{v}) \equiv \mathbf{v}^t \left[\begin{array}{cc} \frac{\partial^2 f}{\partial x^2}(p) & \frac{\partial^2 f}{\partial x \partial y}(p) \\ \frac{\partial^2 f}{\partial x \partial y}(p) & \frac{\partial^2 f}{\partial y^2}(p) \end{array} \right] \mathbf{v}.$$

For example, if

$$\mathbf{v} = \left[\begin{array}{c} 1 \\ 0 \end{array} \right], \qquad \kappa = \frac{\partial^2 f}{\partial x^2}.$$

The bilinear form $(D^2 f)_p$ on $T_p S$ is called the *second fundamental form* II of S at p, given in coordinates as a symmetric 2×2 matrix:

$$\text{II} = D^2 f = \left[\begin{array}{cc} \frac{\partial^2 f}{\partial x^2} & \frac{\partial^2 f}{\partial x \partial y} \\ \frac{\partial^2 f}{\partial x \partial y} & \frac{\partial^2 f}{\partial y^2} \end{array} \right].$$

This formula is good only at the point where the surface is tangent to the x,y-plane. For the second fundamental form, we will always use orthonormal coordinates.

Since II is symmetric, we may choose coordinates x, y such that II is diagonal:

$$\text{II} = \begin{bmatrix} \kappa_1 & 0 \\ 0 & \kappa_2 \end{bmatrix}.$$

Then the curvature κ in the direction $\mathbf{v} = (\cos\theta, \sin\theta)$ is given by Euler's formula (1760):

$$\kappa = \text{II}(\mathbf{v}, \mathbf{v}) = \mathbf{v}^T \text{II} \mathbf{v} = \kappa_1 \cos^2\theta + \kappa_2 \sin^2\theta,$$

a weighted average of κ_1 and κ_2. In particular, the largest and smallest curvatures are κ_1 and κ_2, obtained in the orthogonal directions we have chosen for the x- and y-axes. By the way, Leonard Euler is featured on a ten-franc Swiss note (Figure 3.2).

3.1. Definitions. At a point p in a surface $S \subset \mathbf{R}^3$, the eigenvalues κ_1, κ_2 of the second fundamental form II are called the *principal curvatures*, and the corresponding eigenvectors (uniquely determined unless $\kappa_1 = \kappa_2$) are called the *principal directions* or directions of curvature. The trace of II, $\kappa_1 + \kappa_2$, is called the *mean curvature H*. The determinant of II, $\kappa_1 \kappa_2$, is called the *Gauss curvature G*.

Note that the signs of II and H but not of G depend on the choice of unit normal $\mathbf{n}$. Some treatments define the mean curvature as

$$\frac{1}{2}\,\text{trace II} = \frac{\kappa_1 + \kappa_2}{2}.$$

Just as the curvature κ gave the rate of change of the length of an evolving curve in Chapter 2, the mean curvature H gives the rate of change of the area of an evolving surface. Just as the rate of change of a function of several variables is called the directional derivative and depends on the direction of change, the initial rate of change of the area of a surface depends on its initial velocity $\mathbf{V}$ and is called the first variation.

Figure 3.2. Leonard Euler is featured on this ten-franc Swiss note.

3.2. Theorem. *Let S be a continuous surface in $\mathbf{R}^3$. The first varia-tion of the area on S with respect to a compactly supported continuous vectorfield $\mathbf{V}$ on S is given by integrating $\mathbf{V}$ against the mean curvature:*

$$\delta^1(S) = -\int_S \mathbf{V} \cdot H\mathbf{n}.$$

Remark. $\delta^1(S)$ is defined as

$$\frac{d}{dt} \text{ area } (S + t\mathbf{V})|_{t=0},$$

or, equivalently,

$$\frac{d}{dt} \text{ area } (f_t(S))|_{t=0},$$

where f_t is any C^3 deformation of space with initial velocity $\mathbf{V}$ on S. [$\delta^1(S)$ depends only on $\mathbf{V}$ and is linear in $\mathbf{V}$.] If S has infinite area, restrict attention to the support of $\mathbf{V}$ (where $\mathbf{V} \neq 0$).

Proof. Since the formula is linear in **V**, we may consider tangential and normal variations separately. For tangential variations, which correspond to sliding the surface along itself, $\delta^1(S) = 0$, confirming the formula. Let $V\mathbf{n}$ be a small normal variation, and consider an infinitesimal square area $dx\,dy$ at p, where we may assume the principal directions point along the axes. To first order, the new infinitesimal area is

$$(1 - V\kappa_1)dx(1 - V\kappa_2)dy \approx (1 - VH)dx\,dy = (1 - \mathbf{V}\cdot H\mathbf{n})dx\,dy$$

(compare to Figure 2.2). The formula follows.

Remark. A physical surface such as a soap film would tend to move in the normal direction of positive mean curvature in order to decrease its area, unless balanced by an opposite pressure. The mean curvature of a soap bubble in equilibrium is proportional to the pressure difference across it.

3.3. Minimal surfaces. It follows from Theorem 3.2 that an area-minimizing surface, which minimizes area in competition with surfaces with the same boundary, must have vanishing mean curvature. Any surface with zero mean curvature is called a *minimal surface.*

Some famous minimal surfaces are pictured in Figures 3.3 through 3.6. At each point, since the mean curvature vanishes, the principal curvatures must be equal in magnitude and opposite in sign.

3.4. Coordinates, length, metric. Local coordinates or parameters u_1, u_2 on a C^2 surface $S \subset \mathbf{R}^3$ are provided by a C^2 diffeomorphism (or parameterization) between a domain in the u_1, u_2-plane and a portion of S.

For example, the standard spherical coordinates φ, θ provide local coordinates on all of the sphere of radius a except for the poles (where the longitude θ is undefined and φ is not differentiable). The position vector determined by these coordinates is

$$\mathbf{x} = (x, y, z) = (a\sin\varphi\cos\theta, a\sin\varphi\sin\theta, a\cos\varphi).$$

In general, the position is some function of the coordinates u_i. Along a curve, these coordinates are in turn functions of a single parameter t.

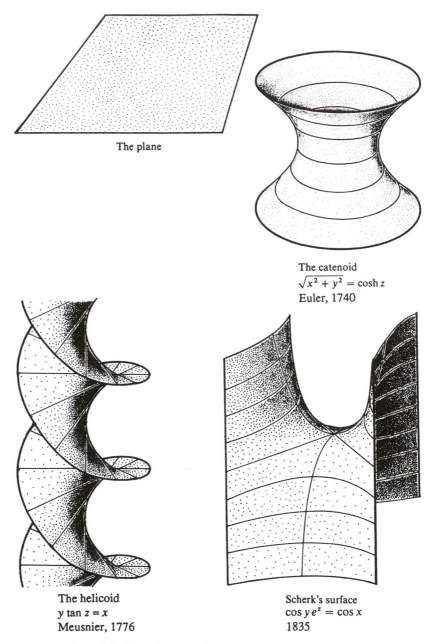

The plane

The catenoid
$$\sqrt{x^2 + y^2} = \cosh z$$
Euler, 1740

The helicoid
$y \tan z = x$
Meusnier, 1776

Scherk's surface
$\cos y\, e^z = \cos x$
1835

Figure 3.3. Some famous minimal surfaces. (Morgan [Mor4], p. 68. All rights reserved. Reprinted with permission of the publisher.)

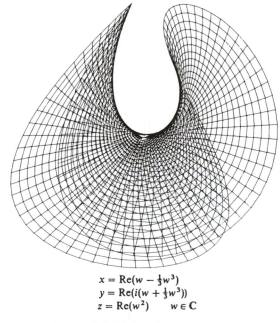

$$x = \text{Re}(w - \tfrac{1}{3}w^3)$$
$$y = \text{Re}(i(w + \tfrac{1}{3}w^3))$$
$$z = \text{Re}(w^2) \qquad w \in \mathbf{C}$$

Figure 3.4. Enneper's surface 1864.

Subscripts on the position vector $\mathbf{x}$ will denote partial derivatives with respect to the u_i:

$$\mathbf{x}_i = \frac{\partial \mathbf{x}}{\partial u_i} = \left(\frac{\partial x}{\partial u_i}, \frac{\partial y}{\partial u_i}, \frac{\partial z}{\partial u_i} \right).$$

A dot will denote differentiation with respect to t:

$$\dot{\mathbf{x}} = \frac{d\mathbf{x}}{dt} = \sum \mathbf{x}_i \dot{u}_i \qquad (\text{the chain rule}).$$

The arc length of a curve in the surface with coordinates $u(t)$ is given by

$$L = \int_{t_0}^{t_1} |\dot{\mathbf{x}}| dt = \int_{t_0}^{t_1} |\mathbf{x}_1 \dot{u}_1 + \mathbf{x}_2 \dot{u}_2| dt$$

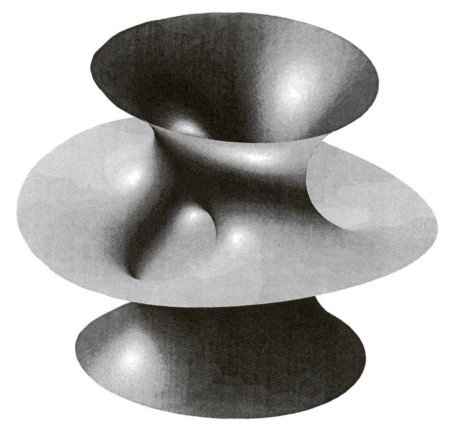

Figure 3.5. The first modern, complete, embedded minimal surface of Costa, Hoffman, and Meeks [Cos], [Hof1], [Hof2]. (Courtesy of David Hoffmann, Jim Hoffmann, and Michael Callahan.)

$$= \int_{t_0}^{t_1} \sqrt{(\mathbf{x}_1 \cdot \mathbf{x}_1)\dot{u}_1^2 + 2(\mathbf{x}_1 \cdot \mathbf{x}_2)\dot{u}_1\dot{u}_2 + (\mathbf{x}_2 \cdot \mathbf{x}_2)\dot{u}_2^2}\,dt \quad (3.1)$$

$$= \int_{t_0}^{t_1} \sqrt{\sum g_{ij}\dot{u}_i\dot{u}_j}\,dt,$$

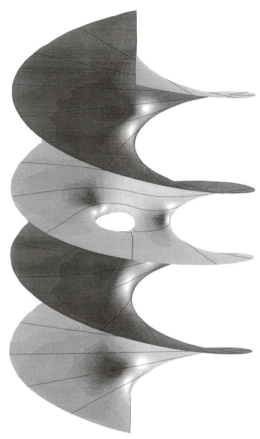

Figure 3.6. One of the latest complete, embedded minimal sur-
faces: the genus one helicoid, discovered by David Hoffman,
Fusheng Wei, and Hermann Karcher [Hof3](1993). Computer-
generated image made by James T. Hoffman at the GANG
Laboratory, University of Massachusetts, Amherst. Copyright
GANG, 1993.

where

$$g_{ij} = \mathbf{x}_i \cdot \mathbf{x}_j = \frac{\partial \mathbf{x}}{\partial u_i} \cdot \frac{\partial \mathbf{x}}{\partial u_j}. \tag{3.2}$$

In other words, $L = \int ds$, where

$$ds^2 = \sum g_{ij} du_i du_j. \tag{3.3}$$

For example, on the sphere of radius a, $L = \int ds$, where, as it turns out,

$$ds^2 = a^2 d\varphi^2 + a^2 \sin^2 \varphi \, d\theta^2 = (a^2 \dot\varphi^2 + a^2 \sin^2 \varphi \, \dot\theta^2) dt^2,$$

so $g_{11} = a^2$, $g_{22} = a^2 \sin^2 \varphi$, and $g_{12} = g_{21} = 0$ (see Exercise 3.4).

The matrix $g = [g_{ij}]$ is called the *first fundamental form* or *metric*. It is an intrinsic quantity in that it relates to measurements inside the surface. Notice that in the formula for length,

$$\begin{aligned}
\sum g_{ij} \dot u_i \dot u_j &= g_{11} \dot u_1 \dot u_1 + g_{12} \dot u_1 \dot u_2 + g_{21} \dot u_2 \dot u_1 + g_{22} \dot u_2 \dot u_2 \\
&= g_{11} \dot u_1^2 + 2 g_{12} \dot u_1 \dot u_2 + g_{22} \dot u_2^2.
\end{aligned}$$

For many surfaces in $\mathbf{R}^3$, it is convenient to use x, y as local coordinates and consider $z(x, y)$. Then

$$\mathbf{x}_1 = (1, 0, z_x) \qquad \text{and} \qquad \mathbf{x}_2 = (0, 1, z_y).$$

The following proposition gives useful formulas for the mean curvature H and Gauss curvature G. g^{-1} denotes the inverse of the first fundamental form matrix $[g_{ij}]$.

3.5. Proposition. *For any local coordinates u_1, u_2 about a point p in a C^2 surface in $\mathbf{R}^3$, the second fundamental form II at p is similar to*

$$(g^{-1})(D^2 \mathbf{x}) \cdot \mathbf{n} = g^{-1} \begin{bmatrix} \mathbf{x}_{11} \cdot \mathbf{n} & \mathbf{x}_{12} \cdot \mathbf{n} \\ \mathbf{x}_{12} \cdot \mathbf{n} & \mathbf{x}_{22} \cdot \mathbf{n} \end{bmatrix},$$

where

$$\mathbf{x}_{ij} = \frac{\partial^2 \mathbf{x}}{\partial u_i \partial u_j} \qquad \text{and} \qquad \mathbf{n} = \frac{\mathbf{x}_1 \times \mathbf{x}_2}{|\mathbf{x}_1 \times \mathbf{x}_2|}.$$

Consequently,

$$\begin{aligned}
H &= \text{trace } g^{-1}(D^2 \mathbf{x}) \cdot \mathbf{n} \tag{3.4} \\
&= \frac{\mathbf{x}_2^2 \mathbf{x}_{11} - 2(\mathbf{x}_1 \cdot \mathbf{x}_2) \mathbf{x}_{12} + \mathbf{x}_1^2 \mathbf{x}_{22}}{\mathbf{x}_1^2 \mathbf{x}_2^2 - (\mathbf{x}_1 \cdot \mathbf{x}_2)^2} \cdot \mathbf{n},
\end{aligned}$$

$$G = \det(g^{-1}(D^2\mathbf{x}) \cdot \mathbf{n}) \tag{3.5}$$

$$= \frac{(\mathbf{x}_{11} \cdot \mathbf{n})(\mathbf{x}_{22} \cdot \mathbf{n}) - (\mathbf{x}_{12} \cdot \mathbf{n})^2}{\mathbf{x}_1^2 \mathbf{x}_2^2 - (\mathbf{x}_1 \cdot \mathbf{x}_2)^2}.$$

Before turning to the proof of Proposition 3.5, we note that

(A) Given H and G, you can solve easily for the principal curvatures:

$$\kappa = \frac{H \pm \sqrt{H^2 - 4G}}{2}.$$

(B) If the surface is a graph $\mathbf{x} = (x, y, f(x, y))$, then

$$H = \frac{(1 + f_y^2)f_{xx} - 2f_x f_y f_{xy} + (1 + f_x^2)f_{yy}}{(1 + f_x^2 + f_y^2)^{3/2}}, \tag{3.6}$$

$$G = \frac{f_{xx}f_{yy} - f_{xy}^2}{(1 + f_x^2 + f_y^2)^2}. \tag{3.7}$$

Proof. We may move S tangent to the x,y-plane at $p = 0$, so S is locally a graph $z = f(x, y)$ with $f_x(0) = f_y(0) = 0$ and $\mathbf{n}(0) = (0, 0, 1)$. For the particular local coordinates x, y,

$$\mathbf{x} = (x, y, f(x, y)), \qquad g(0) = I.$$

The proposition says that II is similar to

$$\begin{bmatrix} f_{xx} & f_{xy} \\ f_{xy} & f_{yy} \end{bmatrix}_0,$$

which is correct; indeed, they are equal.

Now let u_1, u_2 be any local coordinates, and let J denote the Jacobian at 0:

$$J = \begin{bmatrix} \frac{\partial x}{\partial u_1} & \frac{\partial x}{\partial u_2} \\ \frac{\partial y}{\partial u_1} & \frac{\partial y}{\partial u_2} \end{bmatrix}_0.$$

Since

$$\frac{\partial z}{\partial u_1}\Big|_0 = \frac{\partial z}{\partial u_2}\Big|_0 = 0,$$

$$g = J^T J.$$

Then by the chain rule, since $\frac{\partial \mathbf{x}}{\partial x} \cdot \mathbf{n} = 0$ and $\frac{\partial \mathbf{x}}{\partial y} \cdot \mathbf{n} = 0$,

$$(g^{-1})(D^2\mathbf{x}) \cdot \mathbf{n} = (J^T J)^{-1} J^T \begin{bmatrix} \frac{\partial^2 \mathbf{x}}{\partial x^2} \cdot \mathbf{n} & \frac{\partial^2 \mathbf{x}}{\partial x \partial y} \cdot \mathbf{n} \\ \frac{\partial^2 \mathbf{x}}{\partial x \partial y} \cdot \mathbf{n} & \frac{\partial^2 \mathbf{x}}{\partial y^2} \cdot \mathbf{n} \end{bmatrix} J = J^{-1}\mathrm{II}J$$

is indeed similar to II.

Example. We will compute the curvature of the catenoid $\sqrt{x^2 + y^2} = \cosh z$ of Figure 3.2. At most points we could use x and y as coordinates. Instead, we will use z and the polar coordinate θ. The equation says that $r = \cosh z$. Hence the position

$$
\begin{aligned}
\mathbf{x} &= (x, y, z) = (\cosh z \cos\theta, \cosh z \sin\theta, z) \\
\mathbf{x}_1 &= (\sinh z \cos\theta, \sinh z \sin\theta, 1) \\
\mathbf{x}_2 &= (-\cosh z \sin\theta, \cosh z \cos\theta, 0) \\
\mathbf{x}_{11} &= (\cosh z \cos\theta, \cosh z \sin\theta, 0) \\
\mathbf{x}_{12} &= (-\sinh z \sin\theta, \sinh z \cos\theta, 0) \\
\mathbf{x}_{22} &= (-\cosh z \cos\theta, -\cosh z \sin\theta, 0) \\
\mathbf{n} &= \frac{(\cos\theta, \sin\theta, \sinh z)}{\cosh z}.
\end{aligned}
$$

By Proposition 3.5,

$$H = \frac{\cosh^2 z\, \mathbf{x}_{11} - 0 + \cosh^2 z\, \mathbf{x}_{22}}{\text{something}} \cdot \mathbf{n} = 0.$$

so the catenoid is indeed a minimal surface and $\kappa_1 = -\kappa_2$.

$$G = \frac{(1)(-1) - 0}{\cosh^2 z \cosh^2 z - 0} = -\cosh^{-4} z.$$

Hence $\kappa_1 = -\kappa_2 = \cosh^{-2} z$.

3.6. Gauss's Theorema Egregium. *Gauss curvature G is intrinsic. Specifically, there are local coordinates u_1, u_2 about any point p in a C^3 surface S in $\mathbf{R}^3$ such that the first fundamental form g at p is I to first order. In any such coordinate system, the Gauss curvature is*

$$G = \frac{\partial^2 g_{12}}{\partial u_1 \partial u_2} - \frac{1}{2}\frac{\partial^2 g_{22}}{\partial u_1^2} - \frac{1}{2}\frac{\partial^2 g_{11}}{\partial u_2^2}.$$

Remark. To say that $G = I$ to first order means that $g_{11}(p) = g_{22}(p) = 1, g_{12}(p) = 0$, and each

$$g_{ij,k}(p) = \frac{\partial g_{ij}}{\partial u_k}(p) = 0.$$

Proof. Locally S is the graph of a function f over its tangent plane. Orthonormal coordinates on the tangent plane make the metric g equal to I to first order. We may assume that S is tangent to the x, y-plane at p. In x, y coordinates,

$$g = \begin{bmatrix} 1 + f_x^2 & f_x f_y \\ f_x f_y & 1 + f_y^2 \end{bmatrix},$$

and one computes that at p

$$\frac{\partial^2 g_{12}}{\partial x \partial y} - \frac{1}{2}\frac{\partial^2 g_{22}}{\partial x^2} - \frac{1}{2}\frac{\partial^2 g_{11}}{\partial y^2} = f_{xx}f_{yy} - f_{xy}^2 = \det D^2 f = \det \mathrm{II} = G.$$

Any coordinates for which the metric at p is I to first order agree with orthonormal coordinates on the tangent plane to first order and hence yield the same result.

Remark. A full appreciation of Gauss's Theorema Egregium requires the realization that most extrinsic quantities are not intrinsic.

In $\mathbf{R}^3$, a flat piece of plane can be rolled, or "bent," into a piece of cylinder of radius r without changing anything a bug on the surface could detect. This bending, however, does change the curvature κ of an arc of latitude of the cylinder from 0 in the plane to $1/r$ on the cylinder; does change the principal curvatures κ_1, κ_2 from $0, 0$ to $1/r, 0$; and does

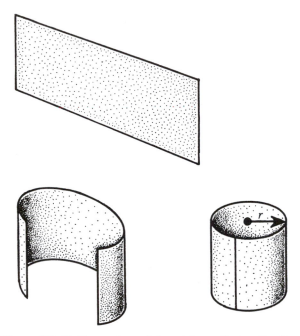

Figure 3.7. Rolling a piece of plane into a cylinder of radius r changes the principal curvatures κ_1, κ_2 from $0, 0$ to $1/r, 0$ and changes the mean curvature $H = \kappa_1 + \kappa_2$ from 0 to $1/r$. The Gauss curvature, however, remains $G = \kappa_1 \kappa_2 = 0$, as Gauss's Theorema Egregium guarantees.

change the mean curvature H from 0 to $1/r$. The Gauss curvature, however, remains $G = \kappa_1 \kappa_2 = 0$, as the theorem guarantees. See Figure 3.7.

No kind of curvature can be detected by a bug on a curve. But if the bug moves to a surface, it can detect Gaussian curvature. By the way, Carl Friedrich Gauss is featured on a ten-mark German note (Figure 3.8).

3.7. Gauss curvature and area. An intrinsic definition of the Gauss curvature G at a point p in a surface could be based on the formula for

Figure 3.8. Carl Friedrich Gauss is featured on this ten-mark German note.

the area of a disc of intrinsic radius r about p:

$$\text{area} = \pi r^2 - G\frac{\pi}{12}r^4 + \cdots . \tag{3.8}$$

(For higher dimensions, see Equation (6.10).) Other interpretations of G will appear in Sections 8.1 and 8.6.

EXERCISES

3.1. Consider the graph $S = \{z = xy\}$ in $\mathbf{R}^3$, tangent to the x, y- plane at the origin. Compute directly the curvature at the origin for fixed direction θ by switching to polar coordinates (r, θ) and taking the second derivative with respect to r. Find the principal directions in which the curvature is greatest and least. What are the principal curvatures κ_1, κ_2; the mean curvature $H = \kappa_1 + \kappa_2$; and the Gauss curvature $G = \kappa_1\kappa_2$ of S at the origin? Let κ_1', κ_2' be the curvature of the directions along the x- and y- axes. What is $\kappa_1' + \kappa_2'$? $\kappa_1'\kappa_2'$?

3.2. What are the principal curvatures κ_1, κ_2; the Gauss curvature G; and the mean curvature H at each of the following?

    a. At a point on a sphere of radius a?

    b. At the origin for the graph $z = f(x, y) = ax^2 + by^2$?

    c. At the origin for the graph $z = f(x, y) = 66x^2 - 24xy + 59y^2$?

    d. At the origin for the graph $z = f(x, y) = x + 2x^2 + 3y^2 + 5y^3$?

    e. At a general point on the helicoid $y \tan z = x$?

    f. At a general point on the ellipsoid $9x^2 + 4y^2 + z^2 = 36$?

3.3. Consider the cylinder $\{x^2 + y^2 = a^2\} \subset \mathbf{R}^3$ with coordinates $u_1 = \theta, u_2 = z$.

    a. Find the parameterization $\mathbf{x}(\theta, z)$; i.e., find $x, y,$ and z as functions of θ and z.

    b. Use Equation (3.2) to find the first fundamental form $[g_{ij}]$ and conclude that $ds^2 = a^2 d\theta^2 + dz^2$.

    c. Use the first fundamental form to compute the length of the helix $z = (h/2\pi)\theta$ $(0 \le \theta \le 2\pi)$, which you can parametrize as $\theta = t$, $z = (h/2\pi)t$. Check that you get the same answer as on page 8.

3.4. For the standard coordinates $u_1 = \theta, u_2 = \varphi$ on the sphere of radius a, compute the first fundamental form $[g_{ij}]$. Use it to calculate the length of a circle of latitude $\varphi = c$.

3.5. Verify that the helicoid is a minimal surface by using Formula (3.6) to compute $H = 0$ for the graph of $f(x, y) = x \tan y$.

3.6. Derive Formulas (3.6) and (3.7) for the curvatures of a graph from Equations (3.4) and (3.5).

3.7. Obtain a surface of revolution in $\mathbf{R}^3$ by revolving a curve $x = f(z)$ in the x, z-plane about the z-axis. Check that the surface is parametrized by cylindrical coordinates θ, z as

$$\mathbf{x} = (f(z)\cos\theta, f(z)\sin\theta, z).$$

Use Proposition 3.5 to show that the inward mean curvature is given by

$$H = \kappa + \frac{1}{f\sqrt{1 + f'^2}}, \qquad (*)$$

where κ is the inward curvature of the original curve ($\kappa = -(f''/(1+ f'^2)^{3/2})$. Check formula $(*)$ for a sphere centered at the origin.

3.8. Check the the catenoid is a minimal surface by using Exercise 3.7 to compute $H = 0$ for $f(z) = \cosh z$.

3.9. A flat planar disk $\{(x, y, z) \in \mathbf{R}^3 : x^2 + y^2 \leq 1, z = 0\}$ moves with initial velocity

$$\mathbf{V} = \begin{cases} (\frac{1}{4} - x^2 - y^2)\mathbf{k} & \text{for } x^2 + y^2 \leq \frac{1}{4}, \\ 0 & \text{for } x^2 + y^2 \geq \frac{1}{4}. \end{cases}$$

What is the initial rate of change of area?

3.10. Do the first-year calculus exercise in spherical coordinates of computing the area of a polar cap of intrinsic radius r on a sphere of radius a to obtain

$$\text{area} = 2\pi a^2 (1 - \cos\frac{r}{a}).$$

Then verify Equation (3.8) for this case.

Surfaces in $\mathbf{R}^n$

This chapter shows how the theory of curvature at a point p in a two-dimensional surface S extends from $\mathbf{R}^3$ to $\mathbf{R}^n$. As before, choose orthonormal coordinates on $\mathbf{R}^n$ with the origin at p and S tangent to the x_1, x_2-plane at p. The tangent plane $T_p S$ to S at p is now the x_1, x_2-plane; the orthogonal complement $T_p S^\perp$ is the $x_3, \cdots, x_n$-plane; and locally S is the graph of a function

$$f : T_p S \to T_p S^\perp.$$

Any unit vector v tangent to S at p, together with the vectors normal to S at p, spans a hyperplane, which intersects S in a curve. The curvature vector κ of this curve, which we call the curvature in the direction v, is just the second derivative

$$\kappa = (D^2 f)_p(v, v).$$

(We will soon have so many tangent vectors v around that we are now abandoning boldface notation **v** for them.)

The bilinear form $(D^2 f)_p$ on $T_p S$ with values in $T_p S^\perp$ is called the *second fundamental tensor* II of S at p, given in coordinates as a symmetric 2×2 matrix with entries in $T_p S^\perp$:

$$\text{II} = \begin{bmatrix} \frac{\partial^2 f}{\partial x_1^2} & \frac{\partial^2 f}{\partial x_1 \partial x_2} \\ \frac{\partial^2 f}{\partial x_1 \partial x_2} & \frac{\partial^2 f}{\partial x_2^2} \end{bmatrix} = \begin{bmatrix} a_{11} & a_{12} \\ a_{12} & a_{22} \end{bmatrix}.$$

Again, this formula is good only at the point where the surface is tangent to the x_1, x_2-plane. If $n = 3$, this second fundamental tensor is just the second fundamental form times the unit normal $\mathbf{n}$.

Generally this matrix cannot be diagonalized to produce principal curvatures. The trace of II, $a_{11} + a_{22} \in T_p S^{\perp}$, is called the *mean curvature vector* $\mathbf{H}$. (If $n = 3$, $\mathbf{H} = H\mathbf{n}$.) The scalar quantity $a_{11} \cdot a_{22} - a_{12} \cdot a_{12}$ is called the *Gauss curvature G.* Neither $\mathbf{H}$ nor G depends on the choice of orthonormal coordinates.

Note that if for example $n = 5$ and $\mathbf{H} = (2, 3, 4)$ as an element of $T_p S^{\perp} \cong \mathbf{R}^3$, then $\mathbf{H} = (0, 0, 2, 3, 4)$ as an element of $\mathbf{R}^5$.

Let G_i denote the Gauss curvature of the projection S_i of S into

$$\mathbf{R}_1 \times \mathbf{R}_2 \times \{0\} \times \cdots \times \mathbf{R}_i \times \cdots \times \{0\} \equiv \mathbf{R}_i^3.$$

Then the Gauss curvature G of S at p is

$$G = \sum_{i=3}^{n} G_i$$

simply because the dot product of two vectors is just the sum of the products of the components.

4.1. Theorem. *Let S be a continuous surface in $\mathbf{R}^n$. The first variation of the area of S with respect to a compactly supported continuous vectorfield $\mathbf{V}$ on S is given by integrating $\mathbf{V}$ against the mean curvature vector:*

$$\delta^1(S) = -\int_S \mathbf{V} \cdot \mathbf{H}.$$

Proof. Since the formula is linear in $\mathbf{V}$, we may consider variations in the $x_1, x_2, \ldots$ directions separately. For the x_1, x_2 directions, which correspond to sliding the surface along itself, $\delta^1(S) = 0$, as the formula says. Let $\mathbf{V}$ be a small variation in the x_3 direction, and consider an infinitesimal square area $dx_1 dx_2$ at p, where we may assume that the x_3 component of II is diagonal:

$$\begin{bmatrix} \kappa_1 & 0 \\ 0 & \kappa_2 \end{bmatrix}.$$

To first order, it is displaced to an infinitesimal area

$$(1 \mp |\mathbf{V}|\kappa_1)dx_1(1 \mp |\mathbf{V}|\kappa_2)dx_2 \approx (1 - \mathbf{V} \cdot \mathbf{H})dx_1 dx_2.$$

The formula follows.

The concepts of local coordinates and the first fundamental form extend without change from $\mathbf{R}^2$ to $\mathbf{R}^n$. Likewise Proposition 3.5 generalizes as follows.

4.2. Proposition. *For any local coordinates u_1, u_2 about a point p in a C^2 surface S in $\mathbf{R}^n$, the second fundamental tensor II at p is similar to*

$$(g^{-1})P(D^2\mathbf{x}) = g^{-1}\left[\begin{array}{cc} P(\mathbf{x}_{11}) & P(\mathbf{x}_{12}) \\ P(\mathbf{x}_{12}) & P(\mathbf{x}_{22}) \end{array}\right],$$

where P denotes projection onto $T_pS^\perp$ and

$$\mathbf{x}_{ij} = \frac{\partial^2 \mathbf{x}}{\partial u_i \partial u_j}.$$

Consequently,

$$\begin{aligned} \mathbf{H} &= \text{trace } g^{-1}P(D^2\mathbf{x}) & (4.1) \\ &= P\frac{\mathbf{x}_2^2 \mathbf{x}_{11} - 2(\mathbf{x}_1 \cdot \mathbf{x}_2)\mathbf{x}_{12} + \mathbf{x}_1^2 \mathbf{x}_{22}}{\mathbf{x}_1^2 \mathbf{x}_2^2 - (\mathbf{x}_1 \cdot \mathbf{x}_2)^2} \end{aligned}$$

$$\begin{aligned} G &= \det\left(g^{-1}P(D^2\mathbf{x})\right) & (4.2) \\ &= \frac{(P\mathbf{x}_{11}) \cdot (P\mathbf{x}_{22}) - (P\mathbf{x}_{12})^2}{\mathbf{x}_1^2 \mathbf{x}_2^2 - (\mathbf{x}_1 \cdot \mathbf{x}_2)^2} \end{aligned}$$

Remark. $T_pS^\perp$ and hence P change from point to point. If $T_pS^\perp$ is just the $x_3, \ldots x_n$-plane, then

$$P(a_1, a_2, a_3, a_4, \ldots) = (0, 0, a_3, a_4, \ldots).$$

If $\mathbf{x}_1, \mathbf{x}_2$ give an *orthogonal* basis for T_pS, then

$$P(\mathbf{w}) = \mathbf{w} - \frac{\mathbf{w} \cdot \mathbf{x}_1}{\mathbf{x}_1 \cdot \mathbf{x}_1}\mathbf{x}_1 - \frac{\mathbf{w} \cdot \mathbf{x}_2}{\mathbf{x}_2 \cdot \mathbf{x}_2}\mathbf{x}_2. \qquad (4.3)$$

If $\mathbf{x}_1, \mathbf{x}_2$ are not orthogonal, compute P by replacing $\mathbf{x}_2$ by

$$\mathbf{x}_2 - \frac{\mathbf{x}_2 \cdot \mathbf{x}_1}{\mathbf{x}_1 \cdot \mathbf{x}_1} \mathbf{x}_1.$$

Example. Consider the surface

$$\{(w, z) \in \mathbf{C}^2 : w = e^z\}.$$

We will use $x = Re\, z$ and $y = Im\, z$ as coordinates. Then

$$
\begin{aligned}
\mathbf{x} &= (x, y, e^x \cos y, e^x \sin y) \\
\mathbf{x}_1 &= (1, 0, e^x \cos y, e^x \sin y) \\
\mathbf{x}_2 &= (0, 1, -e^x \sin y, e^x \cos y) \\
\mathbf{x}_{11} &= (0, 0, e^x \cos y, e^x \sin y) \\
\mathbf{x}_{12} &= (0, 0, -e^x \sin y, e^x \cos y) \\
\mathbf{x}_{22} &= (0, 0, -e^x \cos y, -e^x \sin y)
\end{aligned}
$$

Note that $\mathbf{x}_1^2 = \mathbf{x}_2^2 = 1 + e^{2x}, \mathbf{x}_1 \cdot \mathbf{x}_2 = 0$. Hence,

$$\mathbf{H} = P\frac{(0, 0, 0, 0)}{(1 + e^{2x})^2} = \mathbf{0}.$$

This is a minimal surface. (As a matter of fact, every complex analytic variety is a minimal surface. See Exercise 4.4.)

To compute G, first compute $P(\mathbf{x}_{ij})$. Since $\mathbf{x}_1, \mathbf{x}_2$ are orthogonal,

$$
\begin{aligned}
P(\mathbf{x}_{11}) &= \mathbf{x}_{11} - \frac{\mathbf{x}_{11} \cdot \mathbf{x}_1}{\mathbf{x}_1 \cdot \mathbf{x}_1} \mathbf{x}_1 - \frac{\mathbf{x}_{11} \cdot \mathbf{x}_2}{\mathbf{x}_2 \cdot \mathbf{x}_2} \mathbf{x}_2 \\
&= \frac{(-e^{2x}, 0, e^x \cos y, e^x \sin y)}{1 + e^{2x}}.
\end{aligned}
$$

Similarly,

$$
\begin{aligned}
P(\mathbf{x}_{12}) &= \frac{(0, -e^{2x}, -e^x \sin y, e^x \cos y)}{1 + e^{2x}}, \\
P(\mathbf{x}_{22}) &= \frac{(e^{2x}, 0, -e^x \cos y, -e^x \sin y)}{1 + e^{2x}}.
\end{aligned}
$$

Hence,

$$G = \frac{(-e^{4x} - e^{2x}) - (e^{4x} + e^{2x})}{(1 + e^{2x})^2[(1 + e^{2x})^2 - 0]} = -\frac{2e^{2x}}{(1 + e^{2x})^3}.$$

4.3. Gauss's Theorema Egregium. Finally, Gauss's Theorema Egregium, with the same proof as in Section 3.6, says that G is intrinsic, given in local coordinates u_1, u_2 in which $g = I$ to first order by the formula

$$G = \frac{\partial^2 g_{12}}{\partial u_1 \partial u_2} - \frac{1}{2}\frac{\partial^2 g_{22}}{\partial u_1^2} - \frac{1}{2}\frac{\partial g_{11}}{\partial u_2^2}.$$

EXERCISES

4.1. Find the projection of $(0, 0, 2, 0) \in \mathbf{R}^4$ onto the plane T with basis $(1, 0, 2, 2)$, $(0, 1, -2, 2)$ and onto $T^\perp$. Hint: Make those basis vectors orthonormal first.

4.2. Compute the mean curvature vector and the Gauss curvature at each of the following:

a. At the origin for the graph

$$(z, w) = f(x, y) = (x^2 + 2y^2, 66x^2 - 24xy + 59y^2).$$

Then compare with Exercise 3.2 **b, c.**

b. At a general point of $\{(z, w) \in \mathbf{C}^2 : w = z^2\}$.

4.3. Consider the torus $T^2 = \mathbf{S}^1 \times \mathbf{S}^1 \subset \mathbf{R}^4$.

a. Write T^2 at $(0, 1, 0, 1)$ locally as the graph of a function $(y_1, y_2) = f(x_1, x_2)$.

b. Compute the second fundamental tensor, the mean curvature H, and the Gauss curvature G at $(0, 1, 0, 1)$. (By symmetry, the mean and Gauss curvatures are the same at all points.)

4.4. Show that for the graph of a complex analytic function f,

$$\{w = f(z)\} \subset \mathbf{C}^2,$$

$$\mathbf{H} = \mathbf{0},$$

and $$G = -2|f''(z)|^2(1 + |f'(z)|^2)^{-3}.$$

In particular, the graph of a complex analytic function is a minimal surface. (Compare with the example after Proposition 4.2.)

4.5. *Minimal surface equation.* Show that the graph of a function $f : \mathbf{R}^2 \to \mathbf{R}^{n-2}$ is a minimal surface if and only if

$$(1 + |f_y|^2)f_{xx} - 2(f_x \cdot f_y)f_{xy} + (1 + |f_x|^2)f_{yy} = 0.$$

Compare with Formula (3.6) for the special case $n = 3$.

CHAPTER 5

m-Dimensional Surfaces in $\mathbf{R}^n$

This chapter extends the theory to C^2 m-dimensional surfaces S in $\mathbf{R}^n$. As before, choose orthonormal coordinates on $\mathbf{R}^n$ with the origin at p and S tangent to the $x_1, x_2, \dots, x_m$-plane at p. Locally S is the graph of a function

$$f : T_p S \to T_p S^\perp.$$

A unit vector v tangent to S at p, together with the vectors normal to S at p, spans a plane, which intersects S in a curve. The curvature vector κ of the curve, which we call the curvature in the direction v, is just the second derivative

$$\kappa = (D^2 f)_p(v, v).$$

The bilinear form $(D^2 f)_p$ on $T_p S$ with values in $T_p S^\perp$ is called the *second fundamental tensor* II of S at p, given in coordinates as a symmetric $m \times m$ matrix with entries in $T_p S^\perp$:

$$\text{II} = \begin{bmatrix} \frac{\partial^2 f}{\partial x_1^2} & \cdots & \frac{\partial^2 f}{\partial x_1 \partial x_m} \\ \frac{\partial^2 f}{\partial x_1 \partial x_m} & & \frac{\partial^2 f}{\partial x_m^2} \end{bmatrix}.$$

The trace of II is called the *mean curvature vector* **H**. [Some treatments define **H** as (trace II)/m.]

Hypersurfaces. For hypersurfaces $(n = m + 1)$, II is just the unit normal **n** times a scalar matrix, called the second fundamental form and also denoted by II. $\mathbf{H} = H\mathbf{n}$, where H is the *(scalar) mean curvature*. If we choose coordinates to make the second fundamental form diagonal,

$$\text{II} = \begin{bmatrix} \kappa_1 & & 0 \\ & \ddots & \\ 0 & & \kappa_m \end{bmatrix},$$

then $H = \kappa_1 + \cdots + \kappa_m$. If the unit normal $\mathbf{n} = (n_1, \dots, n_n)$ is extended locally as a unit vectorfield, then $\partial n_n/\partial x_n = 0$, while for $1 \le i \le n-1, \partial n_i/\partial x_i = -\kappa_i$ [compare to Equation (2.2)]. Hence

$$H = -\sum_{i=1}^{n} \partial n_i/\partial x_i \equiv -\text{div } \mathbf{n}.$$

If the hypersurface is given as a level set $\{f(x_1, \dots, x_n) = c\}$ then $\mathbf{n} = \nabla f/|\nabla f|$, where $\nabla f \equiv (\partial f/\partial x_1, \dots, \partial f/\partial x_n)$, and

$$H = -\text{div } \frac{\nabla f}{|\nabla f|}. \tag{5.1}$$

5.1. Theorem. *Let S be a C^2 m-dimensional surface in $\mathbf{R}^n$. The first variation of the area of S with respect to a compactly supported continuous vectorfield $\mathbf{V}$ on S is given by integrating $\mathbf{V}$ against the mean curvature vector:*

$$\delta^1(S) = -\int_S \mathbf{V} \cdot \mathbf{H}.$$

Proof. Since the formula is linear in $\mathbf{V}$, we may consider variations in the $x_1, x_2, \dots, x_n$ directions separately. For the $x_1, \dots, x_m$ directions, which correspond to sliding the surface along itself, $\delta^1(S) = 0$, as the formula says. Let $\mathbf{V}$ be a small variation in the x_j direction $(m < j \le n)$, and consider an infinitesimal area $dx_1 \cdots dx_m$ at p, where we may assume that the x_j component of II is diagonal:

$$\begin{bmatrix} \kappa_1 & & 0 \\ & \ddots & \\ 0 & & \kappa_m \end{bmatrix}.$$

To first order, it is displaced to an infinitesimal area

$$(1 \mp |\mathbf{V}|\kappa_1)dx_1 \cdots (1 \mp |\mathbf{V}|\kappa_m)dx_m \approx (1 - \mathbf{V} \cdot \mathbf{H}) \, dx_1 \cdot dx_m.$$

The formula follows.

5.2. Multi-vectors and the geometry of subspaces of $\mathbf{R}^n$ (Morgan [Mor4]). This section describes a way to represent an m-dimensional subspace of $\mathbf{R}^n$ as a "wedge product" of its basis vectors. Consider $\mathbf{R}^n$ with basis $e_1, e_2, \cdots, e_n$. There is a nice way of multiplying m vectors in $\mathbf{R}^n$ to obtain a new object called an *m-vector* ξ:

$$\xi = v_1 \wedge \cdots \wedge v_m.$$

This wedge product is charcterized by two properties. First, it is multilinear:

$$
\begin{aligned}
cv_1 \wedge v_2 &= v_1 \wedge cv_2 = c(v_1 \wedge v_2), \\
(u_1 + v_1) \wedge (u_2 + v_2) &= u_1 \wedge u_2 + u_1 \wedge v_2 + v_1 \wedge u_2 + v_1 \wedge v_2.
\end{aligned}
$$

Second, it is alternating:

$$u \wedge v = -v \wedge u \quad \text{or} \quad u \wedge u = 0.$$

For example,

$$
\begin{aligned}
(2e_1 + 3e_2 - 5e_3) & \wedge (7e_1 - 11e_3) \\
&= 14e_1 \wedge e_1 - 22e_1 \wedge e_3 + 21e_2 \wedge e_1 - 33e_2 \wedge e_3 \\
&\quad -35e_3 \wedge e_1 + 55e_3 \wedge e_3 \\
&= 0 - 22e_1 \wedge e_3 - 21e_1 \wedge e_2 - 33e_2 \wedge e_3 + 35e_1 \wedge e_3 + 0 \\
&= -21e_{12} + 13e_{13} - 33e_{23}.
\end{aligned}
$$

We have abbreviated e_{12} for $e_1 \wedge e_2$.

In general, computation of $\xi = v_1 \wedge \cdots \wedge v_m$ yields an answer of the form

$$\xi = \sum_{i_1 < \cdots < i_m} a_{i_1 \cdots i_m} e_{i_1 \cdots i_m}.$$

The set of all linear combinations of $\{e_{i_1 \cdots i_m} : i_1 < \cdots < i_m\}$ is the space $\bigwedge_m \mathbf{R}^n$ of m-vectors, a vector space of dimension $\binom{n}{m}$. It has the inner product for which $\{e_{i_1 \cdots i_m} : i_1 < \cdots < i_m\}$ is an orthonormal basis.

The purpose of an m-vector $\xi = v_1 \wedge \cdots \wedge v_m$ is to represent the oriented m-plane P through $\mathbf{0}$ of which $v_1, \ldots, v_m$ give an oriented basis. Fortunately, the wedge product $\xi' = v_1' \wedge \cdots \wedge v_m'$ of another oriented basis for P turns out to be a positive multiple of ξ. For example, replacing v_1 with $v_1' = \sum c_i v_i$ yields

$$v_1' \wedge v_2 \wedge \cdots \wedge v_m = c_1 v_1 \wedge v_2 \wedge \cdots \wedge v_m.$$

If $v_1, \cdots, v_m$ give an orthonormal basis, then $\xi = v_1 \wedge \cdots \wedge v_m$ has length 1. A product $v_1 \wedge \cdots \wedge v_m$ is 0 if and only if the vectors are linearly dependent. For the case $m = n$,

$$v_1 \wedge \cdots \wedge v_n = \det [v_1, \cdots, v_n] \cdot e_{1 \cdots n}.$$

An m-vector ξ is called *simple* or *decomposable* if it can be written as a single wedge product of vectors. For example, in $\bigwedge_2 \mathbf{R}^4$, $e_{12} + 2e_{13} - e_{23} = (e_1 + e_3) \wedge (e_2 + 2e_3)$ is simple, whereas $e_{12} + e_{34}$ is not (see Exercise 5.6). The oriented m-planes through the origin in $\mathbf{R}^n$ are in one-to-one correspondence with the unit, simple m-vectors in $\bigwedge_m \mathbf{R}^n$.

Incidentally, the geometric relationship between two m-planes in $\mathbf{R}^n$ is given by m angles, which appeared at least as early as 1929 (see Sommerville [Som, Chapter 4, §12]), with beautiful later applications to the geometry of Grassmannians (see Wong [Won]) and to area minimization (see Morgan [Mor1, §2.3]).

5.3. Sectional and Riemannian curvature.

The sectional curvature K of S at p assigns to every 2-plane $P \subset T_p S$ the Gauss curvature of the two-dimensional surface

$$S \cap (P \oplus T_p S^\perp).$$

If v, w give an orthonormal basis for P, then by its definition the sectional curvature is

$$K(P) = \mathrm{II}(v, v) \cdot \mathrm{II}(w, w) - \mathrm{II}(v, w) \cdot \mathrm{II}(v, w). \qquad (5.2)$$

For example, if $\mathrm{II} = [a_{ij}]$ and $P = e_1 \wedge e_2$ is the x_1, x_2-plane, then the sectional curvature is

$$K(P) = \mathrm{II}(e_1, e_1) \cdot \mathrm{II}(e_2, e_2) - \mathrm{II}(e_1, e_2) \cdot \mathrm{II}(e_1, e_2)$$
$$= a_{11} \cdot a_{22} - a_{12} \cdot a_{12}.$$

Remark. For hypersurfaces ($n = m + 1$) for any 2-plane $P = \sum p_{ij} \, e_i \wedge e_j$, if we choose coordinates to make the second fundamental form diagonal,

$$\mathrm{II} = \begin{bmatrix} \kappa_1 & & 0 \\ & \ddots & \\ 0 & & \kappa_m \end{bmatrix},$$

then

$$K(P) = \sum_{1 \le i \le j \le m} p_{ij}^2 \kappa_i \kappa_j.$$

Thus any sectional curvature $K(P)$ is a weighted average of the sectional curvatures $\kappa_i \kappa_j$ of the axis 2-planes $e_i \wedge e_j$.

For $2 < m < n$, $\mathbf{R}^n \cong T_p S \times \mathbf{R}_1 \times \cdots \times \mathbf{R}_{n-m}$, let S_i denote the projection of S into $T_p S \times \mathbf{R}_i$, and let K_i denote its sectional curvature. Then, by Equation 5.2, the sectional curvature K of S satisfies $K = \sum K_i$.

Hence the sectional curvature of an m-dimensional surface S in $\mathbf{R}^n$ may be computed by separately diagonalizing the $n - m$ components of II, taking the appropriate weighted average of products of principal curvatures for each component, and summing over all components.

If $m = n - 1$ then II is a symmetric bilinear form called the second fundamental form. Its eigenvalues $\kappa_1, \ldots, \kappa_m$ are called the principal curvatures. Since $(D^2 f)_p$ is symmetric, in some orthogonal coordinates it is diagonal and f takes the form

$$f = \frac{\kappa_1 x_1^2}{2} + \frac{\kappa_2 x_2^2}{2} + \cdots + \frac{\kappa_m x_m^2}{2} + o(\mathbf{x}^2).$$

In general, if $\mathrm{II} = (a_{ij})$, then Formula (5.2) yields

$$K(P) = \left(\sum a_{ik} v_i v_k \right) \cdot \left(\sum a_{jl} w_j w_l \right) - \left(\sum a_{jk} v_k w_j \right) \cdot \left(\sum a_{il} v_i w_l \right)$$
$$= \sum R_{ijkl} v_i w_j v_k w_l, \tag{5.3}$$

where

$$R_{ijkl} = a_{ik} \cdot a_{jl} - a_{jk} \cdot a_{il} \tag{5.4}$$

are the 2×2 minors of II, corresponding to rows i, j and columns k, l. For example, $R_{1234} = a_{13} \cdot a_{24} - a_{14} \cdot a_{23}$ comes from rows 1, 2 and columns 3, 4 of

$$\text{II} = \begin{bmatrix} a_{11} & a_{12} & a_{13} & a_{14} & \cdots \\ a_{12} & a_{22} & a_{23} & a_{24} & \cdots \\ \vdots & \vdots & \vdots & \vdots & \end{bmatrix}.$$

$R_{1212} = a_{11} \cdot a_{22} - a_{12} \cdot a_{12}$ is the sectional curvature of the x_1, x_2-plane.

R is called the *Riemannian curvature tensor*. Thus, the Riemannian curvature tensor is just the 2×2 minors of the second fundamental tensor. Immediately,

$$R_{jikl} = -R_{ijkl} \text{ and } R_{ijlk} = -R_{ijkl} \tag{5.5}$$

(interchanging two rows or columns changes the sign of the minor), and

$$R_{klij} = R_{ijkl} \tag{5.6}$$

because II is symmetric. One can easily check Bianchi's first identity on permutation of the last three indices:

$$R_{ijkl} + R_{iklj} + R_{iljk} = 0. \tag{5.7}$$

To obtain a definition of R independent of the choice of orthonormal coordinates on T_pS, note that R is the bilinear form $\text{II} \wedge \text{II}$ on $\bigwedge_2 T_pS$. Indeed, if $\{e_i\}$ gives a basis for T_pS, so that $\{e_k \wedge e_l : k < l\}$ gives a basis for $\bigwedge_2 T_pS$, then

$$\text{II} \wedge \text{II}(e_k \wedge e_l) = \text{II}(e_k) \wedge \text{II}(e_l) = \left(\sum a_{rk} e_r \right) \wedge \left(\sum a_{sl} e_s \right),$$

and

$$(e_i \wedge e_j) \cdot \text{II} \wedge \text{II}(e_k \wedge e_l) = a_{ik} \cdot a_{jl} - a_{jk} \cdot a_{il} = R_{ijkl}.$$

As a bilinear form on $\bigwedge_2 T_pS$, R is characterized by the values $\zeta \cdot R(\zeta)$ for unit 2-vectors $\zeta \in \bigwedge_2 T_pS$. Actually R is determined by the sectional curvatures $P \cdot R(P)$ for 2-planes (unit *simple* 2-vectors).

The *Ricci curvature* Ric is a bilinear form on T_pS, defined as a kind of trace of the Riemannian curvature. Just as the trace of a matrix $[c_{ij}]$ is a sum $\sum c_{ii}$ over a repeated subscript, the coordinates R_{jl} of the Ricci curvature are given by

$$R_{jl} = \sum_i R_{ijil}. \tag{5.8}$$

If you think of R_{ijkl} as a matrix of matrices,

$$\begin{bmatrix} [R_{i1k1}] & [R_{i1k2}] & \cdots & [R_{i1km}] \\ \vdots & \vdots & \vdots & \vdots \\ [R_{imk1}] & [R_{imk2}] & \cdots & [R_{imkm}] \end{bmatrix}$$

then R_{jl} is the corresponding matrix of traces, so the definition of Ric as a bilinear form does not really depend on the choice of orthonormal coordinates for T_pS. Its application to e_1 yields the sum of the sectional curvatures of axis planes containing e_1:

$$\begin{aligned} e_1 \cdot \text{Ric}(e_1) &= R_{11} = \sum_i R_{i1i1} = \sum_{i \neq 1} R_{i1i1} \\ &= \sum_{i=2}^m K(e_1 \wedge e_i). \end{aligned}$$

Hence for any orthonormal basis $v_1, \ldots, v_m$ for T_pS,

$$v_1 \cdot \text{Ric}(v_1) = \sum_{i=2}^m K(v_1 \wedge v_i), \tag{5.9}$$

and for any unit $v \in T_pS$,

$$v \cdot \text{Ric}(v) = \frac{m-1}{\text{vol}\, \mathbf{S}^{m-2}} \int_{\substack{w \perp v \\ w \in T_pS}} K(v \wedge w). \tag{5.10}$$

Thus the Ricci curvature has an interpretation as an average of sectional curvatures.

The *scalar* curvature R is defined as the trace of the Ricci curvature:

$$R = \sum_i R_{ii}. \tag{5.11}$$

Hence for any orthonormal basis $v_1, \ldots, v_m$ for $T_p S$,

$$R = 2 \sum_{1 \leq i < j \leq m} K(v_i \wedge v_j) = \frac{m(m-1)}{\mathrm{vol}\mathcal{P}} \int_{P \in \mathcal{P}} K(P) \tag{5.12}$$

where $\mathcal{P}$ is the set of all 2-planes in $T_p S$. Thus the scalar curvature is proportional to the average of all sectional curvatures at a point.

Remark. Historically Ric used to have the opposite sign. Some texts give the Riemannian curvature tensor R_{ijkl} the opposite sign.

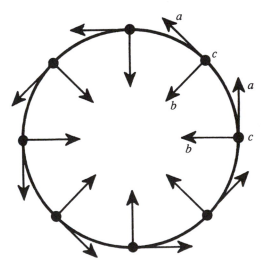

Figure 5.1. (*a*) A vectorfield **f** on the circle, (*b*) its derivative, and (*c*) its covariant derivative, 0.

5.4. The covariant derivative. Let S be a C^2 m-dimensional surface in $\mathbf{R}^n$. If f is a differentiable function on S, then the derivative ∇f is a tangent vectorfield. But if $\mathbf{f}$ is a vectorfield (or a field of matrices or tensors), pointwise in $T_p S$, then the derivative generally will have components normal to S. The projection into $T_p S$ is called the *covariant derivative*. See Figures 5.1 and 5.2 and Exercise 5.11. (The name comes from certain nice transformation properties in a more general setting; see Chapter 6.)

In local coordinates $u_1, \ldots, u_m$ in which $g = I$ to first order at p, the coordinates of the covariant derivative of $\mathbf{f}$ at p are given by the several

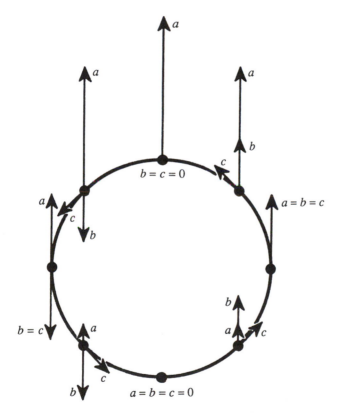

Figure 5.2. (*a*) A vectorfield $\mathbf{f}$ on the circle, (*b*) its derivative, and (*c*) its covariant derivative.

partial derivatives. For example, the coordinates $f_{i;j}$ of the covariant derivative of a vectorfield with coordinates f_i are given by

$$f_{i;j} = f_{i,j} \equiv \frac{\partial f_i}{\partial u_j}.$$

EXERCISES

5.1. Find the mean curvature at the origin of

$$w = f(x, y, z) = 2x^2 + 3y^2 + 4z^2 + 5xz + 6xy + 7yz + 8z^3$$

a. by computing the second fundamental form and

b. as a level surface of $g(x, y, z, w) = w - f(x, y, z)$, using the formula $H = -\text{div}\,(\nabla g / |\nabla g|)$.

5.2. Compute $(e_1 + 2e_2 + 3e_3) \wedge (e_1 + 2e_2 - 3e_3) \wedge e_4$.

5.3. Consider the 2-plane P in $\mathbf{R}^4$ given by

$$P = \{(x_1, x_2, x_3, x_4) : x_1 + x_2 + x_3 = x_3 + x_4 = 0\}.$$

Find a nonorthogonal basis u, v and an orthonormal basis w, z for P. Verify by direct computation that $u \wedge v$ is a multiple of $w \wedge z$ and that $|w \wedge z| = 1$.

5.4. Verify by direct computation that

$$(e_1 + 2e_2 + 3e_3) \wedge (e_1 - e_3) \wedge (e_2 + e_3) = \begin{vmatrix} 1 & 1 & 0 \\ 2 & 0 & 1 \\ 3 & -1 & 1 \end{vmatrix} e_{123}.$$

5.5. Prove that the 2-vector $e_{12} + 2e_{13} + 2e_{23}$ is simple.

5.6. Prove that $e_{12} + e_{34}$ is not simple.

5.7. From the first principles (without computations), explain why the sectional curvatures of an m-dimensional round sphere $\mathbf{S}^m(a)$ of radius a are all $1/a^2$, and hence the scalar curvature $R = m(m - 1)/a^2$. Recall that by Exercise 4.3, for a torus $\mathbf{S}^1(a) \times \mathbf{S}^1(b)$, the Gauss curvature is 0. Use these facts to deduce the various sectional curvatures $K(e_i \wedge e_j)$ of $\mathbf{S}^2(a) \times \mathbf{S}^3(b) \subset \mathbf{R}^3 \times \mathbf{R}^4 = \mathbf{R}^7$ at $p = (0, 0, a;\ 0, 0, 0, b)$ with tangent directions e_1, e_2, e_3, e_4, e_5, and verify in this case the general formula for scalar curvature of a product:

$$R_{S_1 \times S_2}(p) = R_{S_1}(p_1) + R_{S_2}(p_2).$$

5.8. Suppose the second fundamental form at a point p of a three-dimensional surface S in $\mathbf{R}^5$ is given by

$$
\mathrm{II}(p) =
\begin{bmatrix}
\begin{bmatrix} 1 \\ 1 \end{bmatrix} & \begin{bmatrix} 1 \\ 1 \end{bmatrix} & \begin{bmatrix} 0 \\ 2 \end{bmatrix} \\[1em]
\begin{bmatrix} 1 \\ 1 \end{bmatrix} & \begin{bmatrix} 2 \\ 3 \end{bmatrix} & \begin{bmatrix} 1 \\ 0 \end{bmatrix} \\[1em]
\begin{bmatrix} 0 \\ 2 \end{bmatrix} & \begin{bmatrix} 1 \\ 0 \end{bmatrix} & \begin{bmatrix} 0 \\ 0 \end{bmatrix}
\end{bmatrix}
$$

a. Find $K(e_1 \wedge e_2)$ from the definition.

b. Use Equation (5.2) to find $K(P)$ if P has orthonormal basis $(e_1 + e_2)/\sqrt{2}, (e_1 - e_2 + e_3)/\sqrt{3}$.

c. Compute the Riemannian curvature tensor R_{ijkl}. (By Equation (5.5), you may assume $1 \le i < j \le 3, 1 \le k < l \le 3$, so there are just nine components to compute, some of which are equal by Equation (5.6).) Check Bianchi's first identity, Equation (5.7), for $ijkl = 1223$.

d. Use **c** and Equation (5.3) to check **a** and **b**.

5.9. This problem studies the curvature at the origin of the three-dimensional surface in $\mathbf{R}^5$ given by

$$y_1 = x_1^2 + 2x_1x_2 + x_2^2 + 5x_3^2,$$
$$y_2 = 3x_1^2 + x_2^2 + 2x_2x_3 + x_3^2.$$

a. What is II (at the origin)?

b. What is the sectional curvature of the x_1, x_2-plane?

c. What is the sectional curvature of the plane $x_1 + x_2 = 0$? Of the plane $x_1 + x_2 + x_3 = 0$?

d. Give all the components of the Riemannian curvature tensor. Use them to recompute the answers to parts **b** and **c**.

e. Compute the Ricci and scalar curvatures.

5.10. Consider the point $p = (0, 0, 0, -a)$ on the three-dimensional round hypersphere S of radius a about 0 in $\mathbf{R}^4$:

$$S = \{x^2 + y^2 + z^2 + w^2 = a^2\}.$$

a. Find a function $f(x, y, z)$ so that S is the graph of f near p.

b. Show that at p, the second fundamental form II is $1/a$ times the identity matrix. (By symmetry it suffices to compute f_{xx} and f_{xy} at p.)

c. By **b**, what is the scalar mean curvature H at p? The mean curvature vector $\mathbf{H}$ at p?

d. Use **b** to show that every sectional curvature $K(v, w)$ for any orthonormal vectors v, w is $1/a^2$.

e. Use **b** to compute the Riemannian curvature tensor R_{ijkl}.

f. Use **e** to show that every sectional curvature is $1/a^2$.

g. Use **e** to compute the Ricci curvature R_{jl}.

h. Use **g** to compute the scalar curvature R.

i. Now compute H at every point of S by viewing the hypersphere as the level surface of a function $g(x, y, z, w)$ and using the formula $H = -\mathrm{div}(\nabla g / |\nabla g|)$. (You may get minus your previous answer.)

j. For the parameterization of S given by

$$\mathbf{x}(u_1, u_2, u_3) =$$
$$a(\sin u_1 \cos u_2, \sin u_1 \sin u_2, \cos u_1 \cos u_3, \cos u_1 \sin u_3),$$

show that the first fundamental form is

$$ds^2 = a^2 du_1{}^2 + a^2 \sin^2 u_1 du_2{}^2 + a^2 \cos^2 u_1 du_3{}^2.$$

k. A bug's position on S is given by

$$\mathbf{x}(t) = a(\cos t, \sin t, 0, 0) \qquad (0 \le t \le 2\pi).$$

What is the velocity vector $\mathbf{v}$? The unit tangent vector $\mathbf{T}$? The curvature vector κ? The projection of κ onto the tangent plane to the surface? The length of the bug's path? Check the length by using your answer to **j**. (Note that the bug's path is given by $u_1 = \pi/2$, $u_2 = t$, $u_3 = 0$.)

5.11. Consider the vectorfield on $\mathbf{R}^3$: $\mathbf{f} = y^2 \mathbf{i} + (x + z)\mathbf{j} + yz\mathbf{k}$.

a. Compute its derivative at a general point in $\mathbf{R}^3$.

b. Compute its covariant derivative at $(0, 0, 1)$ on the unit sphere.

5.12. Show that for the graph of a function $f : \mathbf{R}^{n-1} \to \mathbf{R}$,

$$H = \mathrm{div} \frac{\nabla f}{\sqrt{1 + |\nabla f|^2}} = \frac{(1 + |\nabla f|^2)\Delta f - \sum f_i f_j f_{ij}}{(1 + |\nabla f|^2)^{3/2}},$$

where

$$f_i = \partial f / \partial x_i,$$
$$f_{ij} = \partial^2 f / \partial x_i \partial x_j,$$
$$\nabla f = (f_1, \dots, f_{n-1}),$$
$$\operatorname{div}(p, q, \dots) = p_1 + q_2 + \cdots,$$

and
$$\triangle f = \operatorname{div} \nabla f = f_{11} + f_{22} + \cdots.$$

CHAPTER 6

Intrinsic Riemannian Geometry

Since many analytic geometric quantities are intrinsic to a smooth m-dimensional surface S in $\mathbf{R}^n$, the standard treatment avoids all reference to an ambient $\mathbf{R}^n$. The surface S is defined as a topological manifold covered by compatible C^∞ coordinate charts, with a "Riemannian metric" g (any smooth positive definite matrix). This is not really a more general setting, since J. Nash [Nas] has proved that every such abstract Riemannian manifold can be isometrically embedded in some $\mathbf{R}^n$. I suppose that it is a more natural setting, but the formulas get much more complicated.

So far we have seen one intrinsic quantity, the Gaussian curvature G of a two-dimensional surface in $\mathbf{R}^n$. We proved G intrinsic by deriving a formula for G in terms of the metric.

One may think of intrinsic Riemannian geometry as nothing but a huge collection of such formulas, thus proving intrinsic such quantities as Riemannian curvature, sectional curvature, and covariant derivatives. The standard approach uses these formulas as definitions. We have the advantage of having the simpler extrinsic definitions behind us. Formulas get much more complicated in intrinsic local coordinates.

In particular, complications arise because the local coordinates fail to be orthogonal, as in Figure 6.1. The u_1-axis is not perpendicular to the level set $\{u_1 = 0\}$; or infinitesimally, the unit vector $e_1 = \partial/\partial u_1$ is not

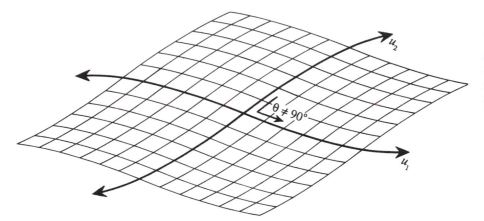

Figure 6.1. For nonorthogonal coordinates, the u_1-axis is not perpendicular to the u_2-axis (the level set $\{u_1 = 0\}$). Infinitesimally, the unit vector $e_1 = \partial/\partial u_1$ is not perpendicular to the level set $\{du_1 = 0\}$.

perpendicular to the level set $\{du_1 = 0\}$. Hence the components of a vectorfield

$$X = (X^1, X^2, \dots, X^m) = \sum X^i e_i = \sum X^i \frac{\partial}{\partial u^i}$$

and the components of a differential

$$\varphi = (\varphi_1, \varphi_2, \dots, \varphi_m) = \sum \varphi_i du^i$$

behave very differently, under changes in coordinates for example. To emphasize the distinction, superscripts are used on the components of vector-like, or *contravariant*, tensors, and subscripts are used on the components of differential-like, or *covariant*, tensors.

Thus a vectorfield X has components X^i. Its covariant derivative has components $X^i_{;j}$, distinguished by the semicolon from the partial derivatives $X^i_{,j} = \partial X^i / \partial u_j$. As our first exercise in intrinsic Riemannian geometry, we will prove that the components of the covariant derivative are given by the formula

$$X^i_{;j} = X^i_{,j} + \sum_k \Gamma^i_{jk} X^k, \tag{6.1}$$

where Γ^i_{jk} are the Christoffel symbols

$$\Gamma^i_{jk} = \frac{1}{2} \sum_l g^{il} (g_{lj,k} + g_{lk,j} - g_{jk,l}), \qquad (6.2)$$

defined in terms of the partial derivatives of the metric g_{ij} and its inverse g^{ij}. In particular, covariant differentiation is an intrinsic notion.

In Formula (6.1), the partial derivative gives the first, main term. There are additional terms because the basis vectors themselves are turning.

In both Formulas (6.1) and (6.2), note how each index i, j, or k on the left occurs in the same position (as a subscript or superscript) on the right. Note how summation runs over the index k or l which appears as both a subscript and a superscript. By these conventions our notation will respect covariance and contravariance.

Some treatments consider covariant differentiation Γ^i_{jk} on manifolds without metrics. Covariant differentiation is also called a *connection*, because by providing for the differentiation of vectorfields it gives some connection between the tangent spaces of S at different points. Our canonical connection which comes from a Riemannian metric (the "Levi-Civita connection") is symmetric, so the *torsion* is 0:

$$T^i_{jk} = \Gamma^i_{jk} - \Gamma^i_{kj} = 0. \qquad (6.3)$$

The Riemannian curvature is given by the formula

$$R^i_{jkl} = -\Gamma^i_{jk,l} + \Gamma^i_{jl,k} + \sum_h (-\Gamma^h_{jk} \Gamma^i_{hl} + \Gamma^h_{jl} \Gamma^i_{hk}). \qquad (6.4)$$

The Riemannian curvature is thus intrinsic, because the connection Γ^i_{jk} is intrinsic. Note that each index on the left occurs in the same position on the right and that summation runs over the index h which appears as both a subscript and a superscript.

The old symmetries (5.5)–(5.7) still hold for the related tensor

$$R_{ijkl} = \sum_h g_{ih} R^h_{jkl}.$$

Since $R^i_{jkl} = \sum_h g^{ih} R_{hjkl}$,

$$R^i_{jlk} = -R^i_{jkl}, \tag{6.5}$$

but in general $R^j_{ikl} \neq -R^i_{jkl}$. For example, R^2_{2kl} need not vanish. The first Bianchi identity still holds:

$$R^i_{jkl} + R^i_{klj} + R^i_{ljk} = 0. \tag{6.6}$$

The Ricci curvature is given by the formula

$$R_{jl} = \sum_i R^i_{jil}, \tag{6.7}$$

and the scalar curvature by the formula

$$R = \sum g^{jl} R_{jl}. \tag{6.8}$$

The sectional curvature of a plane with orthonormal basis v, w is given by

$$K(v \wedge w) = \sum_{i,j,k,l} R_{ijkl} v^i w^j v^k w^l. \tag{6.9}$$

If S is two-dimensional, its Gauss curvature is $G = R/2$.

Note that if $g = I$ to first order at p, then

$$
\begin{aligned}
\Gamma^i_{jk} &= 0, \\
X^i_{:j} &= X^i_{,j}, \\
R_{ijkl} &= R^i_{jkl} = -\Gamma^i_{jk,l} + \Gamma^i_{jl,k}, \\
R_{jl} &= \sum_i R^i_{jil}, \\
R &= \sum R_{jj},
\end{aligned}
$$

and

$$K(v \wedge w) = \sum R^i_{jkl} v^i w^j v^k w^l.$$

Remark. An intrinsic definition of the scalar curvature R at a point p in an m-dimensional surface S could be based on the formula for the volume of a ball of intrinsic radius r about p:

$$\text{volume} = \alpha_m r^m - \alpha_m \frac{r}{6(m+2)} r^{m+2} + \cdots , \tag{6.10}$$

where α_m is the volume of a unit ball in $\mathbf{R}^m$. When $m = 2$, this formula reduces to Equation (3.8). The analogous formula for spheres played a role in R. Schoen's solution of the Yamabe problem of finding a conformal deformation of a given Riemannian metric to one of constant scalar curvature (see Schoen [Sch, Lemma 2]).

6.1. More useful formulas. There are a few more special formulas needed sometimes. The covariant derivative of a general tensor f is given by the formula

$$
f^{j_1 \cdots j_s}_{i_1 \cdots i_r;k} = f^{j_1 \cdots j_s}_{i_1 \cdots i_r,k} + \sum_m \Gamma^{j_1}_{mk} f^{mj_2 \cdots j_s}_{i_1 \cdots i_r} + \cdots
$$
$$
+ \sum_m \Gamma^{j_s}_{mk} f^{j_1 \cdots j_{s-1} m}_{i_1 \cdots i_r} - \sum_m \Gamma^m_{ki_1} f^{j_1 \cdots j_s}_{mi_2 \cdots i_r} - \cdots
$$
$$
- \sum_m \Gamma^m_{ki_r} f^{j_1 \cdots j_s}_{i_1 \cdots i_{r-1} m}. \tag{6.11}
$$

Ricci's lemma says that the covariant derivative of the metric is 0:

$$g_{ij;k} = g^{ij}_{;k} = 0. \tag{6.12}$$

In general, the mixed partial covariant derivatives of a vectorfield X are not equal. *Ricci's identity* gives a very nice formula for the difference in terms of the Riemannian curvature:

$$X^i_{;j;k} - X^i_{;k;j} = - \sum_h R^i_{hjk} X^h. \tag{6.13}$$

Ricci's identity thus provides an alternative description of the Riemannian curvature as a failure of equality for mixed partials. In intrinsic formulations of Riemannian geometry, Ricci's identity is sometimes turned into a definition of Riemannian curvature.

6.2. Proofs. There are two ways to prove the intrinsic formulas of Riemannian geometry: either directly from the extrinsic definitions or more intrinsically by exploiting the invariance under changes of coordinates. As an example, we prove the formula for the covariant derivative of a vectorfield both ways and then compare the two methods.

Extrinsic proof of Equation (6.1). Consider a differentiable vectorfield

$$X = \sum_i X^i \frac{\partial}{\partial u^i} = \sum_{k,m} X^k \frac{\partial x^m}{\partial u^k} \frac{\partial}{\partial x^m}.$$

The ordinary partial derivative satisfies

$$\frac{\partial X}{\partial u^j} = \sum_i X^i_{,j} \frac{\partial}{\partial u^i} + \sum_{k,m} X^k \frac{\partial^2 x^m}{\partial u^j \partial u^k} \frac{\partial}{\partial x^m} \tag{6.14}$$

by the product rule. To obtain the covariant derivative, we project the derivative onto $T_p S$, the span of $x_1, x_2, \ldots, x_m$, the column space or range of the matrix

$$A = \left[\frac{\partial x^i}{\partial u^j} \right].$$

It is a well-known fact from linear algebra that the projection matrix is

$$P = A(A^t A)^{-1} A^t = A g^{-1} A^t.$$

(A is generally not square, but $A^t A$ is square and invertible.)

In Equation (6.14), the first summation over i already lies in $T_p S$. We consider the second summation over k, m. To get the coefficient of $\partial/\partial x^n$ in the projection, we multiply the coefficient of $\partial/\partial x^m$ in Equation (6.14) by the n, m-entry of $P = A g^{-1} A^t$, which is

$$\sum_{i,l} \frac{\partial x^n}{\partial u^i} g^{il} \frac{\partial x^m}{\partial u^l},$$

to obtain

$$\sum \frac{\partial x^n}{\partial u^i} g^{il} \frac{\partial x^m}{\partial u^l} \frac{\partial^2 x^m}{\partial u^j \partial u^k} X^k \frac{\partial}{\partial x^n} = \sum_i \left(\sum_l g^{il} x_{jk} \cdot x_l \right) X^k \frac{\partial}{\partial u^i}.$$

Therefore

$$X^i_{;j} = X^i_{,j} + \sum_k \Gamma^i_{jk} X^k,$$

where

$$\begin{aligned}
\Gamma^i_{jk} &= \sum_l g^{il} \mathbf{x}_{jk} \cdot \mathbf{x}_l \\
&= \sum_l g^{il} (\tfrac{1}{2})[(\mathbf{x}_l \cdot \mathbf{x}_j)_k + (\mathbf{x}_l \cdot \mathbf{x}_k)_j - (\mathbf{x}_j \cdot \mathbf{x}_k)_l] \\
&= \frac{1}{2} \sum_l g^{il} (g_{lj,k} + g_{lk,j} - g_{jk,l}).
\end{aligned}$$

Notice how at the final steps we passed from extrinsic quantities $\mathbf{x}_{jk}$ to the intrinsic $g_{rs,t}$.

Invariance proof of Equation (6.1). In this method of proof, we first check the formula in coordinates for which $g = I$ to first order at p and then check that the formula is invariant under changes of coordinates.

If $g = I$ to first order, Equation (6.1) says that $X^i_{;j} = X^i_{,j}$, which we accept after a few moments' reflection.

To check invariance, we must show that if $u^i, u^{i\prime}$ give coordinates at p, then

$$X^{i\prime}_{;j} = \frac{\partial u^{i\prime}}{\partial u^m} \frac{\partial u^n}{\partial u^{j\prime}} X^m_{;n},$$

where we henceforth agree to sum over repeated indices. (Getting such formulas right—knowing whether the $\partial u^{i\prime}$ goes on the top or the bottom—is perhaps the hardest part of linear algebra, but our index conventions make it automatic.) This verification is something of a mess, but here we go. First we note that

$$g'_{jl} = \frac{\partial u^r}{\partial u^{j\prime}} \frac{\partial u^s}{\partial u^{l\prime}} g_{rs}.$$

Hence

$$g'_{jl,k} = \frac{\partial u^r}{\partial u^{j\prime}} \frac{\partial u^s}{\partial u^{l\prime}} \frac{\partial u^t}{\partial u^{k\prime}} g_{rs,t} + \left[\frac{\partial^2 u^r}{\partial u^{j\prime} \partial u^{k\prime}} \frac{\partial u^s}{\partial u^{l\prime}} + \frac{\partial u^r}{\partial u^{j\prime}} \frac{\partial^2 u^s}{\partial u^{l\prime} \partial u^{k\prime}} \right] g_{rs},$$

and similar equations hold for $-g_{jk,l}, g_{kl,j}$. Combining all three with the definition of $\Gamma^{i\prime}_{jk}$ and

$$g^{il\prime} = \frac{\partial u^{i\prime}}{\partial u^p} \frac{\partial u^{l\prime}}{\partial u^q} g^{pq}$$

yields

$$\Gamma^{i\prime}_{jk} = \frac{1}{2} \frac{\partial u^{i\prime}}{\partial u^p} \frac{\partial u^{l\prime}}{\partial u^q} g^{pq} \left(\frac{\partial u^r}{\partial u^{j\prime}} \frac{\partial u^s}{\partial u^{l\prime}} \frac{\partial u^t}{\partial u^{k\prime}} [g_{rs,t} - g_{rt,s} + g_{ts,r}] + \Omega \right),$$

where

$$
\begin{aligned}
\Omega &= g_{rs} \left[\frac{\partial^2 u^r}{\partial u^{j\prime} \partial u^{k\prime}} \frac{\partial u^s}{\partial u^{l\prime}} + \frac{\partial u^r}{\partial u^{j\prime}} \frac{\partial^2 u^s}{\partial u^{l\prime} \partial u^{k\prime}} - \frac{\partial^2 u^r}{\partial u^{j\prime} \partial u^{l\prime}} \frac{\partial u^s}{\partial u^{k\prime}} \right. \\
&\qquad \left. - \frac{\partial u^r}{\partial u^{j\prime}} \frac{\partial^2 u^s}{\partial u^{k\prime} \partial u^{l\prime}} + \frac{\partial^2 u^r}{\partial u^{k\prime} \partial u^{j\prime}} \frac{\partial u^s}{\partial u^{l\prime}} + \frac{\partial u^r}{\partial u^{k\prime}} \frac{\partial^2 u^s}{\partial u^{j\prime} \partial u^{l\prime}} \right] \\
&= 2g_{rs} \left[\frac{\partial^2 u^r}{\partial u^{j\prime} \partial u^{k\prime}} \frac{\partial u^s}{\partial u^{l\prime}} \right],
\end{aligned}
$$

because g_{rs} is symmetric.

Therefore

$$
\begin{aligned}
\Gamma^{i\prime}_{jk} &= \frac{1}{2} \frac{\partial u^{i\prime}}{\partial u^p} \frac{\partial u^r}{\partial u^{j\prime}} \frac{\partial u^t}{\partial u^{k\prime}} \delta^s_q g^{pq} (g_{rs,t} - g_{rt,s} + g_{ts,r}) \\
&\quad + \frac{\partial^2 u^r}{\partial u^{j\prime} \partial u^{k\prime}} \frac{\partial u^{i\prime}}{\partial u^p} \delta^s_q g^{pq} g_{rs},
\end{aligned}
$$

where $\delta^s_q = 1$ if $s = q$ and 0 otherwise. Since $\delta^s_q g^{pq} g_{rs} = g^{pq} g_{qr} = \delta^p_r$,

$$
\begin{aligned}
\Gamma^{i\prime}_{jk} &= \frac{\partial u^{i\prime}}{\partial u^p} \frac{\partial u^r}{\partial u^{j\prime}} \frac{\partial u^t}{\partial u^{k\prime}} \left[\frac{1}{2} g^{ps} (g_{rs,t} - g_{rt,s} + g_{ts,r}) \right] + \frac{\partial^2 u^r}{\partial u^{j\prime} \partial u^{k\prime}} \frac{\partial u^{i\prime}}{\partial u^r} \\
&= \frac{\partial u^{i\prime}}{\partial u^p} \frac{\partial u^r}{\partial u^{j\prime}} \frac{\partial u^t}{\partial u^{k\prime}} \Gamma^p_{rt} + \frac{\partial^2 u^h}{\partial u^{j\prime} \partial u^{k\prime}} \frac{\partial u^{i\prime}}{\partial u^h},
\end{aligned}
$$

by changing the dummy index in the last term from r to h.

Multiplying both sides by $\partial u^h / \partial u^{i\prime}$ and changing primes and indices yields

$$\frac{\partial^2 u^{i\prime}}{\partial u^m \partial u^l} = \Gamma^h_{lm} \frac{\partial u^{i\prime}}{\partial u^h} - \Gamma^{i\prime}_{hk} \frac{\partial u^{h\prime}}{\partial u^l} \frac{\partial u^{k\prime}}{\partial u^m}. \qquad (6.15)$$

Now

$$X^{i\prime}_{;j} \equiv \frac{\partial}{\partial u^{j\prime}} X^{i\prime} + \Gamma^{i\prime}_{jk} X^{k\prime}$$

$$= \frac{\partial}{\partial u^{j\prime}} \left(\frac{\partial u^{i\prime}}{\partial u^m} X^m \right) + \Gamma^{i\prime}_{jk} \left(\frac{\partial u^{k\prime}}{\partial u^m} X^m \right)$$

$$= X^m_{,n} \frac{\partial u^n}{\partial u^{j\prime}} \frac{\partial u^{i\prime}}{\partial u^m} + X^m \frac{\partial^2 u^{i\prime}}{\partial u^m \partial u^l} \frac{\partial u^l}{\partial u^{j\prime}} + X^m \frac{\partial u^{k\prime}}{\partial u^m} \Gamma^{i\prime}_{jk}.$$

By Equation (6.15),

$$X^{i\prime}_{;j} = X^m_{,n} \frac{\partial u^n}{\partial u^{j\prime}} \frac{\partial u^{i\prime}}{\partial u^m}$$

$$+ X^m \left[\Gamma^h_{lm} \frac{\partial u^{i\prime}}{\partial u^h} \frac{\partial u^l}{\partial u^{j\prime}} - \Gamma^{i\prime}_{hk} \frac{\partial u^{h\prime}}{\partial u^l} \frac{\partial u^{k\prime}}{\partial u^m} \frac{\partial u^l}{\partial u^{j\prime}} + \Gamma^{i\prime}_{jk} \frac{\partial u^{k\prime}}{\partial u^m} \right]$$

$$= \frac{\partial u^{i\prime}}{\partial u^m} \frac{\partial u^n}{\partial u^{j\prime}} X^m_{,n} + \frac{\partial u^{i\prime}}{\partial u^h} \frac{\partial u^l}{\partial u^{j\prime}} \Gamma^h_{lm} X^m.$$

By changing dummy variables in the second term ($m \rightarrow k, h \rightarrow m, l \rightarrow n$), we obtain

$$X^{i\prime}_{;j} = \frac{\partial u^{i\prime}}{\partial u^m} \frac{\partial u^n}{\partial u^{j\prime}} [X^m_{,n} + \Gamma^m_{nk} X^k] = \frac{\partial u^{i\prime}}{\partial u^m} \frac{\partial u^n}{\partial u^{j\prime}} X^m_{;n},$$

as desired.

Remark. Of the two proofs, the first has the advantages of being shorter and deriving the formula, whereas the second proves a given formula.

6.3. Geodesics. Let C be a C^2 curve in a C^2 m-dimensional surface S in $\mathbf{R}^n$, with curvature vector κ at a point $p \in C$. We define the *geodesic curvature* κ_g as the projection of κ onto the tangent space $T_p S$. Equivalently, κ_g is the covariant derivative of the unit tangent vector. While curvature κ is extrinsic, geodesic curvature κ_g is intrinsic.

A *geodesic* is a curve with $\kappa_g = 0$ at all points. For example, geodesics on spheres are arcs of great circles, but other circles of latitude are not geodesics. (See Figure 6.2.) Shortest paths turn out to be geodesics, but there are sometimes also other longer geodesics between pairs of points. For example, nonantipodal points on the equator are joined by a short and a long geodesic, depending on which way you go. The poles are joined by infinitely many semicircular meridians of longitude, all of the same length.

The following theorem explains why smooth shortest paths must be geodesics.

6.4. Theorem. *A smooth curve is a geodesic if and only if the first variation of its length vanishes.*

Proof. Let $\mathbf{x}(t)$ be a local parametrization by arc length. Corresponding to an infinitesimal, compactly supported change $\delta\mathbf{x}$ in $\mathbf{x}(t)$ is a variation

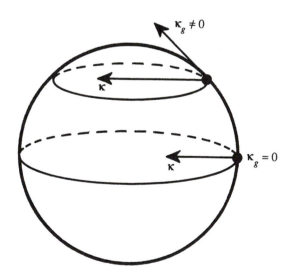

Figure 6.2. On the sphere, great circles are geodesics ($\kappa_g = 0$), but other circles of latitude are not ($\kappa_g \neq 0$).

in length

$$\delta L \;=\; \delta \int (\dot{\mathbf{x}} \cdot \dot{\mathbf{x}})^{1/2} = \int \frac{1}{2}(\dot{\mathbf{x}} \cdot \dot{\mathbf{x}})^{-1/2} 2\dot{\mathbf{x}} \cdot \delta\dot{\mathbf{x}}$$

$$= \int \mathbf{T} \cdot \delta\dot{\mathbf{x}} = -\int \dot{\mathbf{T}} \cdot \delta\mathbf{x} = -\int \boldsymbol{\kappa} \cdot \delta\mathbf{x} = -\int \kappa_g \cdot \delta\mathbf{x}$$

by integration by parts and the fact that $\delta\mathbf{x}$ stays in the surface. δL vanishes for all $\delta\mathbf{x}$ along the surface if and only if $\kappa_g = 0$ and the curve is a geodesic.

Remark. It follows from the theory of differential equations that in any C^3 m-dimensional surface, through any point in any direction there is locally a unique C^2 geodesic. Actually C^2 suffices by work of Philip Hartman ([Har1, §6(i), p. 283] or [Har2, Example 6.2, p. 106]). Such a geodesic provides the shortest path to nearby points. A little more argument shows that if S is connected and compact (or connected and merely complete), between *any* two points some geodesic provides the shortest path (the Hopf-Rinow Theorem, see Cheeger and Ebin [Che1, Chapter 3] or Helgason [Hel, Theorem 10.4]).

6.5. Formula for geodesics. In local coordinates $u^1, \ldots, u^m$, consider a curve $u(t)$ parametrized by arc length, so that the unit tangent vector $\mathbf{T} = \dot{u}$. The derivative of any function $f(u)$ along the curve is given by $\sum f_j \dot{u}^j$ (the chain rule). The covariant derivative of any vectorfield X^i along the curve satisfies

$$\sum_j X^i_{;j} \dot{u}^j \;=\; \sum_j X^i_{,j} \dot{u}^j + \sum_{j,k} \Gamma^i_{jk} \dot{u}^j X^k$$

$$= \dot{X}^i + \sum_{j,k} \Gamma^i_{jk} \dot{u}^j X^k \qquad (6.16)$$

[see 6.1]. Hence for a geodesic (parametrized by arc length), the covariant derivative along the curve of the vectorfield $X^i = T^i = \dot{u}^i$ must vanish:

$$0 = \ddot{u}^i + \sum_{j,k} \Gamma^i_{jk} \dot{u}^j \dot{u}^k. \qquad (6.17)$$

6.6. Hyperbolic geometry. As an example in Riemannian geometry, we consider two-dimensional hyperbolic space H for which global coordinates are given by the upper halfplane

$$\{(x, y) \in \mathbf{R}^2 : y > 0\}$$

with metric

$$g_{ij} = y^{-2} \delta_{ij};$$

that is,

$$ds^2 = \frac{1}{y^2} dx^2 + \frac{1}{y^2} dy^2.$$

Since pointwise g is a multiple of the standard metric (g is "conformal"), angles are the same in the upper halfplane as on H, although distances are different, of course.

Now we compute the Christoffel symbols and curvature.

$$g^{ij} = y^2 \delta^{ij}.$$

By Formula (6.2),

$$\begin{aligned}
\Gamma^1_{12} = \Gamma^1_{21} &= \frac{1}{2} y^2 (g_{12,1} + g_{11,2} - g_{12,1}) \\
&= \frac{1}{2} y^2 \frac{\partial}{\partial y} y^{-2} = -y^{-1}.
\end{aligned}$$

Similarly,

$$-\Gamma^2_{11} = \Gamma^2_{22} = -y^{-1}$$

and the rest are 0.

By Formula (6.4),

$$\begin{aligned}
R^1_{212} &= -\Gamma^1_{21,2} + \Gamma^1_{22,1} + \sum_h (-\Gamma^h_{21} \Gamma^1_{h2} + \Gamma^h_{22} \Gamma^1_{h1}) \\
&= -y^{-2} + 0 + (-y^{-2} + y^{-2}) = -y^{-2}.
\end{aligned}$$

Similarly,

$$
\begin{aligned}
R^2_{121} &= -y^{-2}, & R^1_{111} = R^2_{222} &= 0, \\
R_{11} &= R^1_{111} + R^2_{121} = -y^{-2}, & R_{22} &= -y^{-2} \\
R &= -2, & G = -1. &
\end{aligned}
$$

Thus hyperbolic space H has constant Gaussian curvature -1 and assumes its exalted place with the plane $(G = 0)$ and the sphere $(G = 1)$.

Geodesics parametrized by arc length t must satisfy the Equation (6.17):

$$
\ddot{x} - 2y^{-1}\dot{x}\dot{y} = 0 \qquad \text{and} \qquad \ddot{y} + y^{-1}\dot{x}^2 - y^{-1}\dot{y}^2 = 0.
$$

Let $p = dx/dy$; then

$$
\dot{x} = p\dot{y}, \qquad \ddot{x} = \frac{dp}{dy}\dot{y}^2 + p\ddot{y}.
$$

Substituting for $\ddot{y}$ from the second equation in the first yields

$$
\frac{dp}{dy} = y^{-1}(p^3 + p).
$$

Integrating by partial fractions gives

$$
\frac{dx}{dy} = p = \pm\frac{cy}{\sqrt{1 - c^2y^2}}.
$$

If $c = 0$, we obtain vertical lines as geodesics. Otherwise, letting $c = 1/a$ and integrating yields

$$
(x - b)^2 + y^2 = a^2.
$$

These geodesics are just semicircles centered on the x-axis. See Figure 6.3.

Through any two points there is a unique geodesic, or "straight line," that provides the shortest path between the points. Indeed, Euclid's first four postulates all hold. The notorious fifth postulate fails. Its equivalent statement due to Playfair says that for a given line l and a point p not on

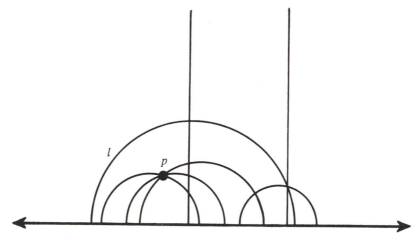

Figure 6.3. Geodesics in hyperbolic space H are semicircles centered on the x-axis and vertical lines.

the line, there is a unique line through p that does not intersect l. The uniqueness fails in hyperbolic geometry, as Figure 6.3 illustrates. Thus hyperbolic geometry proves the impossibility of what geometers had been attempting for millenia—to deduce the fifth postulate from the first four—and gives the premier example of non-Euclidean geometry.

It is interesting to note that the hyperbolic distance from any point (a, b) to the x-axis, measured along a vertical geodesic, is

$$\int_{y=0}^{b} y^{-1} dy = \infty.$$

Hyperbolic space actually has no boundary, but extends infinitely far in all directions.

6.7. Geodesics and sectional curvature. We remark that positive sectional curvature means that parallel geodesics converge, as on the sphere. Negative sectional curvature means that parallel geodesics diverge, as in hyperbolic space.

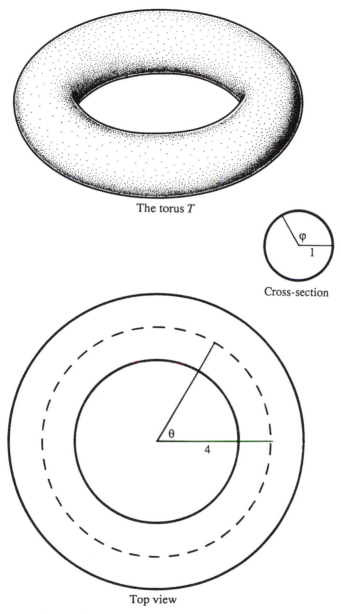

The torus T

Cross-section

Top view

Figure 6.4. The torus T, with coordinates θ, φ.

EXERCISES

6.1. *A torus.* Let T be the torus obtained by revolving a unit circle parametrized by $0 \leq \varphi < 2\pi$ about an axis 4 units from its center. Use the angle $0 \leq \theta < 2\pi$ of revolution and φ as coordinates. (See Figure 6.4.)

a. Show that

$$g_{11} = (4 + \cos \varphi)^2, \qquad g_{12} = g_{21} = 0, \qquad g_{22} = 1,$$

$$\Gamma^1_{12} = -\frac{\sin \varphi}{4 + \cos \varphi}, \qquad \Gamma^2_{11} = \sin \varphi (4 + \cos \varphi),$$

the rest are 0.

b. Consider the vectorfield

$$a = \cos \varphi \frac{\partial}{\partial \theta},$$

that is, $a^1 = \cos \varphi$ and $a^2 = 0$. Show that the covariant derivative is given by

$$a^1_{;2} = -\sin \varphi \frac{4 + 2 \cos \varphi}{4 + \cos \varphi}, \qquad a^2_{;1} = \sin \varphi \cos \varphi (4 + \cos \varphi),$$

the rest are 0.

c. Show that the length of the spiral curve $\theta(t) = \varphi(t) = t \, (0 \leq t \leq 2\pi)$ is given by the integral

$$\int_0^{2\pi} [(4 + \cos t)^2 + 1]^{1/2} dt.$$

d. Of course, $R^1_{111} = R^2_{222} = 0$. Show that

$$R^1_{212} = \frac{\cos \varphi}{4 + \cos \varphi}, \qquad R^2_{121} = \cos \varphi (4 + \cos \varphi),$$

$$R_{11} = \cos \varphi (4 + \cos \varphi), \qquad R_{12} \equiv R_{21} = 0,$$

$$R_{22} = \frac{\cos \varphi}{4 + \cos \varphi}, \qquad R = \frac{2 \cos \varphi}{4 + \cos \varphi}, \qquad G = \frac{\cos \varphi}{4 + \cos \varphi}.$$

e. Conclude that there is no distance-preserving map of any region in the torus on regions in the plane or on the sphere.

6.2. *The flat higher-dimensional torus.* Consider the n-dimensional torus $T^n \subset \mathbf{R}^{2n}$ obtained as the Cartesian product of n circles $\mathbf{S}^1$:

$$T^n = S^1 \times \ldots \times S^1$$
$$= \{(x_1, \cdots, x_{2n}) \in \mathbf{R}^{2n} : x_1^2 + x_2^2 = \ldots = x_{2n-1}^2 + x_{2n}^2 = 1\}.$$

Compute via the second fundamental form and the Christoffel symbols that the sectional curvature of T^n is 0; i.e., $K(P) = 0$ at every point for every 2-plane P. Such a surface is called *flat*.

6.3. *The sphere.* This exercise will verify that geodesics are arcs of great circles and that the sphere has constant Gauss curvature. Consider a sphere S of radius a with the usual spherical coordinates $u^1 = \theta$ and $u^2 = \varphi$.

a. Check that a great circle is given by the equation

$$w = \cot\varphi = c_1 \cos\theta + c_2 \sin\theta,$$

except for vertical great circles $\theta = c$. Hint: A nonvertical great circle is the intersection of a sphere with a plane

$$z = c_1 x + c_2 y.$$

b. Show that the metric is given

$$g_{11} = a^2 \sin^2\varphi, \qquad g_{12} = 0, \qquad g_{22} = a^2.$$

c. Compute the Christoffel symbols

$$\Gamma_{12}^1 = \cot\varphi, \qquad \Gamma_{11}^2 = -\sin\varphi\cos\varphi,$$

the rest vanish. Conclude that geodesics satisfy

$$\begin{cases} \ddot{\theta} + 2\cot\varphi\ \dot{\theta}\dot{\varphi} = 0 \\ \ddot{\varphi} - \sin\varphi\cos\varphi\ \dot{\theta}^2 = 0 \end{cases}$$

and hence $\varphi'' - 2\cot\varphi\varphi'^2 - \sin\varphi\cos\varphi = 0$, where primes denote derivatives with respect to θ, unless $\theta = c$. Thus $w = \cot\varphi$ satisfies $w'' + w = 0$, so

$$w = c_1\cos\theta + c_2\sin\theta.$$

Therefore geodesics are indeed arcs of great circles.

d. Compute the Riemannian curvature

$$R^1_{212} = -R^1_{221} = 1, \qquad R^2_{121} = -R^2_{112} = \sin^2\varphi,$$

the rest are 0; the Ricci curvature

$$R_{11} = \sin^2\varphi, \qquad R_{12} = R_{21} = 0, \qquad R_{22} = 1;$$

the scalar curvature $R = 2a^{-2}$ and finally the Gauss curvature $G = a^{-2}$.

Remark. Actually a simple symmetry argument shows that geodesics are arcs of great circles. We may assume that the geodesic is tangent to the equator at a point. By uniqueness, it must equal its own reflection across the equator. Hence it must be an arc of the equator.

6.4. *The hypersphere.* Consider the three-dimensional round hypersphere S of radius a about 0 in $\mathbf{R}^4$, $\{x^2 + y^2 + z^2 + w^2 = a^2\}$, with parameterization

$$\mathbf{x}(u_1, u_2, u_3) =$$
$$a\left(\sin u_1\cos u_2, \sin u_1\sin u_2, \cos u_1\cos u_3, \cos u_1\sin u_3\right)$$
$$(0 \le u_1 \le \pi/2, 0 \le u_2 \le 2\pi, 0 \le u_3 \le 2\pi).$$

a. Show that the metric is

$$ds^2 = a^2{du_1}^2 + a^2\sin^2 u_1{du_2}^2 + a^2\cos^2 u_1{du_3}^2.$$

b. Compute the Christoffel symbols Γ^i_{jk}. (Note that of all the $g_{ij,k}$, only $g_{22,1}$ and $g_{33,1}$ are nonzero.)

c. Compute some components of the Riemannian curvature tensor:

$$\begin{aligned}
R_{11} &= 2, \\
R_{22} &= 1 - \cos 2u_1 = 2\sin^2 u_1, \\
R_{33} &= 1 + \cos 2u_1 = 2\cos^2 u_1
\end{aligned}$$

d. Find v, a multiple of $\frac{\partial}{\partial u_1}$ of unit length. Use **c** to compute $\text{Ric}(v, v) = R_{jl}v^j v^l$. Explain physically.

e. Use **c** to compute the scalar curvature R. Explain physically.

f. Write down the equations for a geodesic and verify that $u_1 = \pi/4$, $u_2 = u_3 = t$ satisfies them. Double check by computing the curvature vector κ of this curve in $\mathbf{R}^4$ and showing that it is always perpendicular to the sphere S (so that the geodesic curvature is 0). Is this curve a great circle?

6.5. An alternative global model for hyperbolic space is the unit "Poincaré" disc in $\mathbf{R}^2$ with metric $ds = 2ds_0/(1 - r^2)$, where $ds_0^2 = dr^2 + r^2 d\theta^2$, so that

$$ds^2 = \frac{4}{(1 - r^2)^2}dr^2 + \frac{4r^2}{(1 - r^2)^2}d\theta^2.$$

Using coordinates r, θ, compute the Christoffel symbols, Riemannian curvature tensor, scalar curvature R, and finally that the Gauss curvature $R/2 = -1$. Double check the result using this special formula for orthogonal coordinates (Stoker [Sto, Section 13, p. 180]).

$$K = -\frac{1}{2\sqrt{\det g_{ij}}}\left[\frac{\partial}{\partial\theta}\frac{g_{11,2}}{\sqrt{\det g_{ij}}} + \frac{\partial}{\partial r}\frac{g_{22,1}}{\sqrt{\det g_{ij}}}\right].$$

CHAPTER 7

General Relativity

Toward the end of the nineteenth century, a puzzling inconsistency in Mercury's orbit was observed.

Newton had brilliantly explained Kepler's elliptical planetary orbits through solar gravitational attraction and calculus. His successors used a method of perturbations to compute the deviations caused by the other planets. Their calculations predicted that the elliptical orbit shape should rotate, or precess, some fraction of a degree per century:

Planet	Predicted Precession (per century)
Saturn	$46'$
Jupiter	$432''$
Mercury	$532''$

Here $60'$ (60 minutes) equals 1 degree of arc, and $60''$ (60 seconds) equals 1 minute of arc.

Observation confirmed the predictions for Saturn and Jupiter, but showed that Mercury precessed $575''$ per century. By 1900, it was obvious that the variance from the expected precession exceeded any conceivable experimental error. What was causing the additional $43''$ per century?

General relativity would provide the answer.

Figure 7.1. "Mercury's running slow."

7.1. General relativity. The theory of general relativity has three elements. First, special relativity describes motion in free space without gravity. Second, the Principle of Equivalence extends the theory, at least in principle, to include gravity, roughly by equating gravity with acceleration. Third, Riemannian geometry provides a mathematical framework which makes calculations possible.

I first learned the derivation of Mercury's precession from the works of Spain ([Spa, Chapter 8]) and Weinberg ([Wei, Chapter 9]) with the help of my friend Ira Wasserman. The short derivation given here is based on a talk by my student Phat Vu at a mathematics colloquium at Williams College, in turn based on the book of G. B. Jeffery ([Jef, Chapter 7]). A simplified account including some dramatic episodes from the history of astronomy appears in Morgan [Mor2].

7.2. Special relativity. A single particle in free space follows a straight line at constant velocity. For example, $x = at, y = bt, z = ct$, or

$$\frac{x}{a} = \frac{y}{b} = \frac{z}{c},$$

which is the formula for a straight line through the origin in 3-space. This path is also a straight line in four-dimensional space-time:

$$\frac{x}{a} = \frac{y}{b} = \frac{z}{c} = \frac{t}{1},$$

that is, a geodesic for the standard metric

$$ds^2 = dx^2 + dy^2 + dz^2 + dt^2. \tag{7.1}$$

It is actually a geodesic for any metric of the form

$$ds^2 = a_1 dx^2 + a_2 dy^2 + a_3 dz^2 + a_4 dt^2. \tag{7.2}$$

Einstein based special relativity on two axioms:

1. The laws of physics look the same in all inertial frames of reference—that is, to all observers moving with constant velocity relative to one another. (Of course, in accelerating reference frames, the laws of physics look different. Cups of lemonade in accelerating cars suddenly fall over, and tennis balls on the floors of rockets flatten like pancakes.)

2. The speed of a light beam is the same relative to any inertial frame, whether moving in the same or the opposite direction. (Einstein apparently guessed this surprising fact without knowing the evidence provided by the famous Michelson-Morley experiment. It leads to other curiosities, such as time's slowing down at high velocities.)

Einstein's postulates hold for motion along geodesics in space-time if one takes the special case of Equation (7.2):

$$ds^2 = -dx^2 - dy^2 - dz^2 + c^2 dt^2 \qquad (7.3)$$

This is the famous Lorentz metric, with c the speed of light. We will choose units to make $c = 1$.

The Lorentz metric remains invariant under inertial changes of coordinates, but looks funny in accelerating coordinate systems.

For us a new feature of this metric is the presence of minus signs; the metric is not positive definite. Except for the novel fact that the square of the length of curves in space-time can be positive or negative, all definitions and properties remain formally the same (see O'Neill [ONe]). In particular, positive sectional curvature would still imply that parallel geodesics converge (the square of the distance between them decreases). (See Section 6.7.)

This new distance s is often called "proper time" τ, since a motionless particle (x, y, z constant) has $ds^2 = dt^2$. If we replace the symbol s by τ and change to spherical coordinates, the Lorentz metric becomes

$$d\tau^2 = -dr^2 - r^2 d\varphi^2 - r^2 \sin^2 \varphi \, d\theta^2 + dt^2. \qquad (7.4)$$

7.3. The Principle of Equivalence. Special relativity handles motion—position, velocity, acceleration—in free space. The remaining question is how to handle gravity. The Principle of Equivalence asserts that infinitesimally the physical effects of gravity are indistinguishable from those of acceleration. If you feel pressed against the floor of a tiny elevator, you cannot tell whether it is because the elevator is resting on a massive planet or because the elevator is accelerating upward. Consequently, the effect of gravity is just like that of acceleration: it just makes the formula for ds^2 look funny. Computing motion in a gravitational field will reduce to computing geodesics in some strange metric.

7.4. The Schwarzschild metric. The most basic example in general relativity is the effect on the Lorentz metric of a single point mass, such as a sun at the center of a solar system. We will assume that the metric takes the simple form

$$d\tau^2 = -e^{\lambda(r)}dr^2 - r^2(d\varphi^2 + \sin^2\varphi\,d\theta^2) + e^{\nu(r)}dt^2, \qquad (7.5)$$

where $\lambda(r)$ and $\nu(r)$ are functions to be determined. This metric is spherically symmetric and time-independent. (The coordinate r is chosen so that the relativistic distortion is in the radial rather than the tangential directions.) For physical reasons, Einstein further assumed that what is now called the Einstein tensor vanishes:

$$G_k^i = g^{ij}R_{jk} - \frac{1}{2}R\delta_k^i = 0. \qquad (7.6)$$

(Actually, the vanishing of the Einstein tensor G_k^i implies the vanishing of the Ricci curvature R_{jk}, so an equivalent and simpler assumption would be that $R_{jk} = 0$.) To employ this assumption, we now compute the Einstein tensor for the metric (7.5). We order the variables r, φ, θ, t. We compute first the metric

$$g_{11} = -e^\lambda, \quad g_{22} = -r^2, \quad g_{33} = -r^2\sin^2\varphi, \quad g_{44} = e^\nu,$$

others vanish,

$$g^{11} = -e^{-\lambda}, \quad g^{22} = -r^{-2}, \quad g^{33} = -r^{-2}\sin^{-2}\varphi, \quad g^{44} = e^{-\nu},$$

others vanish;

then the Christoffel symbols

$$\Gamma_{11}^1 = \frac{1}{2}\lambda', \quad \Gamma_{22}^1 = -re^{-\lambda}, \quad \Gamma_{33}^1 = -re^{-\lambda}\sin^2\varphi, \qquad (7.7)$$

$$\Gamma_{44}^1 = \frac{1}{2}\nu'e^{\nu-\lambda}, \quad \Gamma_{12}^2 = \Gamma_{13}^3 = r^{-1}, \quad \Gamma_{33}^2 = -\sin\varphi\cos\varphi,$$

$$\Gamma_{23}^3 = \cot\varphi, \quad \Gamma_{14}^4 = \frac{1}{2}\nu',$$

others vanish,

where λ' denotes $d\lambda/dr$; then some components of the Riemannian curvature tensor

$$R^2_{121} = R^3_{131} = \frac{1}{2}r^{-1}\lambda', \quad R^4_{141} = -\frac{1}{2}\nu'' + (\frac{1}{2}\nu')(\frac{1}{2}\lambda') - \frac{1}{4}\nu'^2,$$

$$R^1_{212} = \frac{1}{2}r\lambda'e^{-\lambda}, \quad R^3_{232} = 1 - e^{-\lambda}, \quad R^4_{242} = \frac{1}{2}\nu'(-re^{-\lambda}),$$

$$R^1_{313} = \frac{1}{2}r\lambda'\sin^2\varphi\, e^{-\lambda}, \quad R^2_{323} = \sin^2\varphi\,(1 - e^{-\lambda}),$$

$$R^4_{343} = \frac{1}{2}\nu'(-r)e^{-\lambda}\sin^2\varphi,$$

$$R^1_{414} = \frac{1}{2}e^{\nu-\lambda}(\nu'' + \frac{1}{2}\nu'^2 - \frac{1}{2}\nu'\lambda'),$$

$$R^2_{424} = R^3_{434} = \frac{1}{2}r^{-1}\nu'e^{\nu-\lambda};$$

then some components of the Ricci curvature

$$R_{11} = r^{-1}\lambda' - \frac{1}{2}\nu'' + \frac{1}{4}\nu'\lambda' - \frac{1}{4}\nu'^2,$$

$$R_{22} = 1 + \frac{1}{2}re^{-\lambda}(\lambda' - \nu') - e^{-\lambda},$$

$$R_{33} = \sin^2\varphi\, R_{22},$$

$$R_{44} = \frac{1}{2}e^{\nu-\lambda}(\nu'' + \frac{1}{2}\nu'^2 - \frac{1}{2}\nu'\lambda' + 2r^{-1}\nu'),$$

$$R = -2r^{-2} + e^{-\lambda}(\nu'' - 2r^{-1}\lambda' - \frac{1}{2}\nu'\lambda' + \frac{1}{2}\nu'^2 + 2r^{-1}\nu' + 2r^{-2});$$

and finally some components of the Einstein tensor

$$G^i_k = g^{ij}R_{jk} - \frac{1}{2}R\delta^i_k,$$

$$G^1_1 = r^{-2} + e^{-\lambda}(-r^{-1}\nu' - r^{-2}),$$

$$G^2_2 = G^3_3 = e^{-\lambda}(-\frac{1}{2}\nu'' + \frac{1}{2}r^{-1}\lambda' - \frac{1}{2}r^{-1}\nu' + \frac{1}{4}\nu'\lambda' - \frac{1}{4}\nu'^2),$$

$$G^4_4 = r^{-2} + e^{-\lambda}(r^{-1}\lambda' - r^{-2}).$$

Since $G^4_4 = 0$, $e^{-\lambda} = 1 - \gamma r^{-1}$ for some constant γ (just check that $d\gamma/dr = 0$). Consideration of a test particle with 0 velocity and large r

(see Exercise 7.5) leads to the conclusion that $\gamma = 2GM$, where M is the central mass and G is the gravitational constant. Therefore

$$e^{-\lambda} = 1 - 2GMr^{-1}.$$

Since $G_1^1 = G_4^4 = 0, \lambda + \nu$ is constant. Since the metric should look like the Lorentz metric for r huge, we conclude that $\lambda + \nu = 0$. Therefore

$$e^{\nu} = e^{-\lambda} = 1 - 2GMr^{-1}. \tag{7.8}$$

Now $G_2^2 = G_3^3 = 0$ automatically.

We have obtained the famous Schwarzschild metric

$$\begin{aligned} d\tau^2 &= -(1 - 2GMr^{-1})^{-1}dr^2 - r^2(d\varphi^2 + \sin^2\varphi\,d\theta^2) \\ &+ (1 - 2GMr^{-1})dt^2. \end{aligned} \tag{7.9}$$

Notice that if $M = 0$, the Schwarzschild metric (7.9) reduces to the Lorentz metric (7.4). Notice too the singularity in these coordinates as r decreases to $2GM$, which has this interpretation: shrinking the sun to a point mass has created a black hole of "Schwarzschild radius" $r = 2GM$!

7.5. Relativistic celestial mechanics. Now we are ready to see what differences general relativity predicts for Mercury's orbit. The physics is embodied in the four equations for geodesics, Equation (6.17), in the Schwarzschild metric (7.9). Four equations should let us solve for r, φ, θ, and t as functions of τ. Actually, instead of the first equation for geodesics involving $d^2r/d\tau^2$, we will use the identity $d\tau^2 = g_{ij}dx^idx^j$:

$$-(1 - 2GMr^{-1})^{-1}\left(\frac{dr}{d\tau}\right)^2 - r^2\left(\frac{d\varphi}{d\tau}\right)^2 \tag{7.10}$$

$$-r^2\sin^2\varphi\left(\frac{d\theta}{d\tau}\right)^2 + (1 - 2GMr^{-1})\left(\frac{dt}{d\tau}\right)^2 = 1.$$

To compute the other three geodesic equations, we proceed from Equation (7.8) to compute

$$\begin{aligned} \lambda' &= -\nu' = -2GM(r^2 - 2GMr)^{-1}, \\ \lambda'' &= -\nu'' = 2GM(r^2 - 2GMr)^{-2}(2r - 2GM), \end{aligned}$$

and then, from Equation (7.7),

$$
\begin{aligned}
\Gamma_{11}^1 &= -GM(r^2 - 2GMr)^{-1} \\
\Gamma_{22}^1 &= -r(1 - 2GMr^{-1}) = 2GM - r \\
\Gamma_{33}^1 &= (2GM - r)\sin^2\varphi \\
\Gamma_{44}^1 &= \frac{1}{2}(1 - 2GMr^{-1})(2GMr^{-2}) \\
\Gamma_{12}^2 &= \Gamma_{13}^3 = r^{-1} \\
\Gamma_{33}^2 &= -\sin\varphi\cos\varphi \\
\Gamma_{23}^3 &= \cot\varphi \\
\Gamma_{14}^4 &= GM(r^2 - 2GMr)^{-1}.
\end{aligned}
$$

Hence the last three geodesic equations (see Equation (6.17)) are

$$
\frac{d^2\varphi}{d\tau^2} + 2r^{-1}\frac{dr}{d\tau}\frac{d\varphi}{d\tau} - \sin\varphi\cos\varphi\left(\frac{d\theta}{d\tau}\right)^2 = 0, \qquad (7.11)
$$

$$
\frac{d^2\theta}{d\tau^2} + 2r^{-1}\frac{dr}{d\tau}\frac{d\theta}{d\tau} + 2\cot\varphi\frac{d\varphi}{d\tau}\frac{d\theta}{d\tau} = 0, \qquad (7.12)
$$

$$
\frac{d^2t}{d\tau^2} + \frac{2GM}{r^2 - 2GMr}\frac{dr}{d\tau}\frac{dt}{d\tau} = 0. \qquad (7.13)
$$

The solution of Equations (7.10) through (7.13) will give Mercury's orbit. Assuming that initially $d\varphi/d\tau$ and $\cos\varphi$ are 0, by Equation (7.11) φ remains $\pi/2$. Thus, even relativistically, the orbit remains planar. The other three equations become

$$
-(1 - 2GMr^{-1})^{-1}\left(\frac{dr}{d\tau}\right)^2 - r^2\left(\frac{d\theta}{d\tau}\right)^2
$$

$$
+(1 - 2GMr^{-1})\left(\frac{dt}{d\tau}\right)^2 = 1, \qquad (7.14)
$$

$$
\frac{d^2\theta}{d\tau^2} + 2r^{-1}\frac{dr}{d\tau}\frac{d\theta}{d\tau} = 0, \qquad (7.15)
$$

$$\frac{d^2t}{d\tau^2} + 2GM(r^2 - 2GMr)^{-1}\frac{dr}{d\tau}\frac{dt}{d\tau} = 0. \tag{7.16}$$

Integrating Equation (7.15) and Equation (7.16) yields

$$r^2\frac{d\theta}{d\tau} = h \quad \text{(a constant)}, \tag{7.17}$$

$$(1 - 2GMr^{-1})\frac{dt}{d\tau} = \beta \quad \text{(a constant)}. \tag{7.18}$$

Therefore Equation (7.14) becomes

$$-r^{-4}\left(\frac{dr}{d\theta}\right)^2 - r^{-2}(1 - 2GMr^{-1}) + \beta^2 h^{-2} = h^{-2}(1 - 2GMr^{-1}). \tag{7.19}$$

Putting $r = u^{-1}$ yields

$$\left(\frac{du}{d\theta}\right)^2 = 2GM\left(u^3 - \frac{1}{2GM}u^2 + \beta_1 u + \beta_0\right)$$

for some constants β_0, β_1. The maximum and minimum values u_1, u_2 of u must be roots. Since the roots sum to $1/2GM$, the third root is $1/2GM - u_1 - u_2$, and hence

$$\left(\frac{du}{d\theta}\right)^2 = 2GM(u - u_1)(u - u_2)(u - \frac{1}{2GM} + u_1 + u_2),$$

$$\frac{d\theta}{|du|} = \frac{1}{\sqrt{(u_1 - u)(u - u_2)}}[1 - 2GM(u + u_1 + u_2)]^{-1/2}$$

$$\approx \frac{1 + GM(u + u_1 + u_2)}{\sqrt{(u_1 - u)(u - u_2)}}.$$

To first approximation the orbit is the classical ellipse

$$u = l^{-1}(1 + e\cos\theta),$$

with $u_1 = l^{-1}(1 + e)$, $u_2 = l^{-1}(1 - e)$, and mean distance

$$a = \frac{1}{2}\left(\frac{1}{u_1} + \frac{1}{u_2}\right) = \frac{l}{1 - e^2}.$$

For one revolution,

$$\Delta\theta \approx \int_{\theta=0}^{2\pi} \frac{1 + GMl^{-1}(3 + e\cos\theta)}{\sqrt{l^{-1}e(1 - \cos\theta)l^{-1}e(1 + \cos\theta)}} l^{-1}e|\sin\theta|d\theta$$

$$= \int_{\theta=0}^{2\pi} 1 + GMl^{-1}(3 + e\cos\theta)d\theta$$

$$= 2\pi + 6\pi GM/l$$

$$= 2\pi + 6\pi GM/a(1 - e^2).$$

The ellipse has precessed $6\pi GM/a(1 - e^2)$ radians. The rate of precession in terms of Mercury's period T is

$$\frac{6\pi GM}{a(1 - e^2)T},$$

or, back in more standard units (in which the speed of light c is not 1),

$$\frac{6\pi GM}{c^2a(1 - e^2)T} \text{ radians.}$$

Now

$$\begin{aligned}
G &= \text{gravitational constant} = 6.67 \times 10^{-11} \text{m}^3/\text{kg sec}^2, \\
M &= \text{mass of sun} = 1.99 \times 10^{30} \text{kg}, \\
c &= \text{speed of light} = 3.00 \times 10^8 \text{m/sec}, \\
a &= \text{mean distance from Mercury to sun} = 5.768 \times 10^{10} \text{m}, \\
e &= \text{eccentricity of Mercury's orbit} = 0.206, \\
T &= \text{period of Mercury} = 88.0 \text{ days},
\end{aligned}$$

$$\text{century} \;=\; 36525 \text{ days},$$
$$\text{radian} \;=\; 360/2\pi \text{ degrees},$$
$$\text{degree} \;=\; 3600''.$$

Multiplying these fantastic numbers together, we conclude that the rate of precession is about $43.1''/\text{century}$, in perfect agreement with observation.

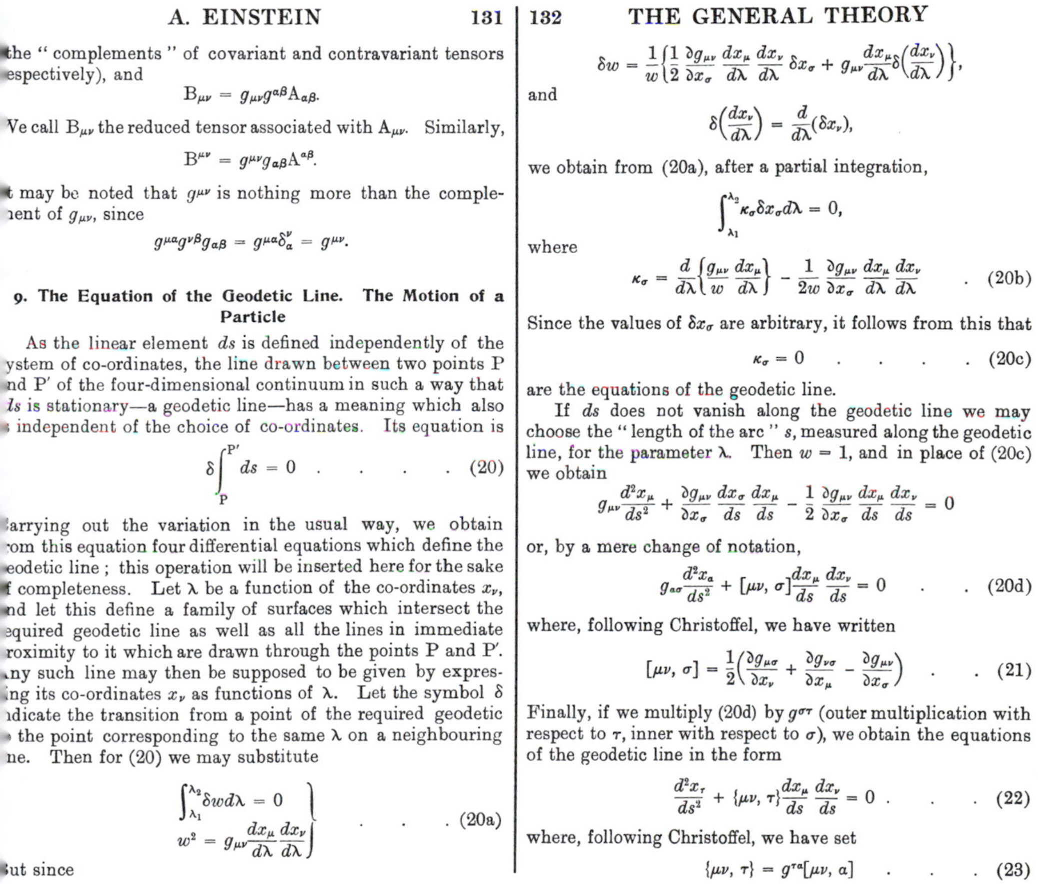

A. EINSTEIN   131

the " complements " of covariant and contravariant tensors respectively), and

$$B_{\mu\nu} = g_{\mu\nu}g^{\alpha\beta}A_{\alpha\beta}.$$

We call $B_{\mu\nu}$ the reduced tensor associated with $A_{\mu\nu}$. Similarly,

$$B^{\mu\nu} = g^{\mu\nu}g_{\alpha\beta}A^{\alpha\beta}.$$

It may be noted that $g^{\mu\nu}$ is nothing more than the complement of $g_{\mu\nu}$, since

$$g^{\mu\alpha}g^{\nu\beta}g_{\alpha\beta} = g^{\mu\alpha}\delta_{\alpha}^{\nu} = g^{\mu\nu}.$$

9. The Equation of the Geodetic Line. The Motion of a Particle

As the linear element ds is defined independently of the system of co-ordinates, the line drawn between two points P and P' of the four-dimensional continuum in such a way that ds is stationary—a geodetic line—has a meaning which also is independent of the choice of co-ordinates. Its equation is

$$\delta \int_{P}^{P'} ds = 0 \qquad . \qquad . \qquad . \qquad (20)$$

Carrying out the variation in the usual way, we obtain from this equation four differential equations which define the geodetic line ; this operation will be inserted here for the sake of completeness. Let λ be a function of the co-ordinates x_ν, and let this define a family of surfaces which intersect the required geodetic line as well as all the lines in immediate proximity to it which are drawn through the points P and P'. Any such line may then be supposed to be given by expressing its co-ordinates x_ν as functions of λ. Let the symbol δ indicate the transition from a point of the required geodetic to the point corresponding to the same λ on a neighbouring line. Then for (20) we may substitute

$$\left.\begin{aligned} \int_{\lambda_1}^{\lambda_2} \delta w \, d\lambda &= 0 \\ w^2 &= g_{\mu\nu}\frac{dx_\mu}{d\lambda}\frac{dx_\nu}{d\lambda} \end{aligned}\right\} \qquad . \qquad . \qquad (20a)$$

But since

---

132   **THE GENERAL THEORY**

$$\delta w = \frac{1}{w}\left\{\frac{1}{2}\frac{\partial g_{\mu\nu}}{\partial x_\sigma}\frac{dx_\mu}{d\lambda}\frac{dx_\nu}{d\lambda}\,\delta x_\sigma + g_{\mu\nu}\frac{dx_\mu}{d\lambda}\delta\left(\frac{dx_\nu}{d\lambda}\right)\right\},$$

and

$$\delta\left(\frac{dx_\nu}{d\lambda}\right) = \frac{d}{d\lambda}(\delta x_\nu),$$

we obtain from (20a), after a partial integration,

$$\int_{\lambda_1}^{\lambda_2} \kappa_\sigma \delta x_\sigma d\lambda = 0,$$

where

$$\kappa_\sigma = \frac{d}{d\lambda}\left\{\frac{g_{\mu\nu}}{w}\frac{dx_\mu}{d\lambda}\right\} - \frac{1}{2w}\frac{\partial g_{\mu\nu}}{\partial x_\sigma}\frac{dx_\mu}{d\lambda}\frac{dx_\nu}{d\lambda} \qquad . \quad (20b)$$

Since the values of δx_σ are arbitrary, it follows from this that

$$\kappa_\sigma = 0 \qquad . \qquad . \qquad . \qquad (20c)$$

are the equations of the geodetic line.

If ds does not vanish along the geodetic line we may choose the " length of the arc " s, measured along the geodetic line, for the parameter λ. Then $w = 1$, and in place of (20c) we obtain

$$g_{\mu\nu}\frac{d^2x_\mu}{ds^2} + \frac{\partial g_{\mu\nu}}{\partial x_\sigma}\frac{dx_\sigma}{ds}\frac{dx_\mu}{ds} - \frac{1}{2}\frac{\partial g_{\mu\nu}}{\partial x_\sigma}\frac{dx_\mu}{ds}\frac{dx_\nu}{ds} = 0$$

or, by a mere change of notation,

$$g_{\alpha\sigma}\frac{d^2x_\alpha}{ds^2} + [\mu\nu,\,\sigma]\frac{dx_\mu}{ds}\frac{dx_\nu}{ds} = 0 \qquad . \qquad (20d)$$

where, following Christoffel, we have written

$$[\mu\nu,\,\sigma] = \frac{1}{2}\left(\frac{\partial g_{\mu\sigma}}{\partial x_\nu} + \frac{\partial g_{\nu\sigma}}{\partial x_\mu} - \frac{\partial g_{\mu\nu}}{\partial x_\sigma}\right) \qquad . \qquad (21)$$

Finally, if we multiply (20d) by $g^{\sigma\tau}$ (outer multiplication with respect to τ, inner with respect to σ), we obtain the equations of the geodetic line in the form

$$\frac{d^2x_\tau}{ds^2} + \{\mu\nu,\,\tau\}\frac{dx_\mu}{ds}\frac{dx_\nu}{ds} = 0 \qquad . \qquad . \qquad (22) \;\Big]$$

where, following Christoffel, we have set

$$\{\mu\nu,\,\tau\} = g^{\tau\alpha}[\mu\nu,\,\alpha] \qquad . \qquad . \qquad (23)$$

Figure 7.2. Einstein's relativity paper [Ein], mostly Riemannian geometry, derives the formula for a geodesic.

If, further, the tensor $A^{\rho\sigma}$ is symmetrical, this reduces to

$$- \tfrac{1}{2}\sqrt{-g}\frac{\partial g_{\rho\sigma}}{\partial x_\mu}A^{\rho\sigma}.$$

Had we introduced, instead of $A^{\rho\sigma}$, the covariant tensor $A_{\rho\sigma} = g_{\rho\alpha}g_{\sigma\beta}A^{\alpha\beta}$, which is also symmetrical, the last term, by virtue of (31), would assume the form

$$\tfrac{1}{2}\sqrt{-g}\frac{\partial g^{\rho\sigma}}{\partial x_\mu}A_{\rho\sigma}.$$

In the case of symmetry in question, (41) may therefore be replaced by the two forms

$$\sqrt{-g}\,A_\mu = \frac{\partial(\sqrt{-g}\,A_\mu^\sigma)}{\partial x_\sigma} - \tfrac{1}{2}\frac{\partial g_{\rho\sigma}}{\partial x_\mu}\sqrt{-g}\,A^{\rho\sigma} \ . \quad (41a)$$

$$\sqrt{-g}\,A_\mu = \frac{\partial(\sqrt{-g}\,A_\mu^\sigma)}{\partial x_\sigma} + \tfrac{1}{2}\frac{\partial g^{\rho\sigma}}{\partial x_\mu}\sqrt{-g}\,A_{\rho\sigma} \ . \quad (41b)$$

which we have to employ later on.

§ 12. The Riemann-Christoffel Tensor

We now seek the tensor which can be obtained from the fundamental tensor *alone*, by differentiation. At first sight the solution seems obvious. We place the fundamental tensor of the $g_{\mu\nu}$ in (27) instead of any given tensor $A_{\mu\nu}$, and thus have a new tensor, namely, the extension of the fundamental tensor. But we easily convince ourselves that this extension vanishes identically. We reach our goal, however, in the following way. In (27) place

$$A_{\mu\nu} = \frac{\partial A_\mu}{\partial x_\nu} - \{\mu\nu, \rho\}A_\rho,$$

i.e. the extension of the four-vector A_μ. Then (with a somewhat different naming of the indices) we get the tensor of the third rank

$$A_{\mu\sigma\tau} = \frac{\partial^2 A_\mu}{\partial x_\sigma \partial x_\tau} - \{\mu\sigma, \rho\}\frac{\partial A_\rho}{\partial x_\tau} - \{\mu\tau, \rho\}\frac{\partial A_\rho}{\partial x_\sigma} - \{\sigma\tau, \rho\}\frac{\partial A_\mu}{\partial x_\rho}$$

$$+ \left[-\frac{\partial}{\partial x_\tau}\{\mu\sigma, \rho\} + \{\mu\tau, a\}\{a\sigma, \rho\} + \{\sigma\tau, a\}\{a\mu, \rho\} \right]A_\rho.$$

This expression suggests forming the tensor $A_{\mu\sigma\tau} - A_{\mu\tau\sigma}$. For, if we do so, the following terms of the expression for $A_{\mu\sigma\tau}$ cancel those of $A_{\mu\tau\sigma}$, the first, the fourth, and the member corresponding to the last term in square brackets; because all these are symmetrical in σ and τ. The same holds good for the sum of the second and third terms. Thus we obtain

$$A_{\mu\sigma\tau} - A_{\mu\tau\sigma} = B^\rho_{\mu\sigma\tau}A_\sigma \quad . \quad . \quad . \quad (42)$$

where

$$B^\rho_{\mu\sigma\tau} = -\frac{\partial}{\partial x_\tau}\{\mu\sigma, \rho\} + \frac{\partial}{\partial x_\sigma}\{\mu\tau, \rho\} - \{\mu\sigma, a\}\{a\tau, \rho\}$$

$$+ \{\mu\tau, a\}\{a\sigma, \rho\} \quad (43)$$

The essential feature of the result is that on the right side of (42) the A_ρ occur alone, without their derivatives. From the tensor character of $A_{\mu\sigma\tau} - A_{\mu\tau\sigma}$ in conjunction with the fact that A_ρ is an arbitrary vector, it follows, by reason of § 7, that $B^\rho_{\mu\sigma\tau}$ is a tensor (the Riemann-Christoffel tensor).

The mathematical importance of this tensor is as follows : If the continuum is of such a nature that there is a co-ordinate system with reference to which the $g_{\mu\nu}$ are constants, then all the $B^\rho_{\mu\sigma\tau}$ vanish. If we choose any new system of co-ordinates in place of the original ones, the $g_{\mu\nu}$ referred thereto will not be constants, but in consequence of its tensor nature, the transformed components of $B^\rho_{\mu\sigma\tau}$ will still vanish in the new system. Thus the vanishing of the Riemann tensor is a necessary condition that, by an appropriate choice of the system of reference, the $g_{\mu\nu}$ may be constants. In our problem this corresponds to the case in which,[*] with a suitable choice of the system of reference, the special theory of relativity holds good for a *finite* region of the continuum.

Contracting (43) with respect to the indices τ and ρ we obtain the covariant tensor of second rank

[*] The mathematicians have proved that this is also a *sufficient* condition.

Figure 7.3. On page 141 appears the formula for the Riemannian curvature tensor, which Einstein calls $B^\rho_{\mu\sigma\tau}$ instead of R^i_{jkl}.

7.6. Einstein's paper. The pages from a translation of Einstein's 1916 paper, "The foundation of the general theory of relativity," [Ein] clearly illustrate that general relativity is basically just Riemannian geometry, which he needed to expound to the world of physicists. The paper begins on page 111. The definition of a metric:

$$ds^2 = \sum_{\tau\sigma} g_{\sigma\tau}dx_\sigma dx_\tau$$

appears on page 119 with Greek instead of our Latin subscripts. The derivation of the formula for a geodesic (using $\{\mu\nu, \tau\}$ instead of Γ^i_{jk} for the Christoffel symbols), from the hypothesis of a vanishing first variation, shows up on pages 131–132 (Figure 7.2). Later, on page 141 (Figure 7.3), he introduces the formula for the Riemannian curvature tensor, denoted $B^\rho_{\mu\sigma\tau}$ instead of R^i_{jkl}. The oversimplifying assumption on page 144 that the Ricci curvature vanishes requires a footnote explaining that the more justifiable assumption that what we now call the Einstein tensor vanishes would lead to the same results (cf. Section 7.4). On page 145 comes this conclusion:

> These equations, which proceed, by the method of pure mathematics, from the requirement of the general theory of relativity, give us, in combination with the equations of motion (46), to a first approximation Newton's law of attraction, and to a second approximation the explanation of the motion of the perihelion of Mercury discovered by Leverrier (as it remains after corrections for perturbation have been made). These facts must, in my opinion, be taken as a convincing proof of the correctness of the theory.

EXERCISES

7.1. What is the precession of the Earth's orbit due to general relativity? ($e \approx .0167, a \approx 1.5 \times 10^{11}$m.) Why do you think this result has not been used to check general relativity?

7.2. A small planet has a perfect circular orbit of radius r_0 about the sun. What is its speed according to general relativity? (Use one of the four equations for geodesics.)

7.3. Einstein was led by cosmological factors to consider the universe with the metric

$$d\tau^2 = -(1 - r^2/R^2)^{-1}dr^2 - r^2(d\varphi^2 + \sin^2\varphi\, d\theta^2) + dt^2.$$

(Here R is a constant, "the radius of the universe.")

a. Is this metric spherically symmetric?

b. What is the elapsed proper time $\Delta\tau$ for a spaceship to circle the origin at $r = r_0$ at $3/5$ the speed of light?

7.4. A space pod, free of the earth, is falling toward the sun from $r = R_1 = 1.5 \times 10^{11}$ meters ($\sim$ Earth's orbit) initially slowly with $dr/dt = (1 - \alpha^2 R_1^{-1})\alpha R_1^{-1/2}$, where $\alpha = \sqrt{2GM}$.

a. According to general relativity, how long will it take it to reach $r = R_2 = 6.96 \times 10^8$ m. ($\sim$ sun's surface), according to a distant observer? (Answer: about $27\frac{1}{2}$ days.)

b. Now assuming the sun has contracted to the "Schwarzschild radius" $R_2 = 2GM = \alpha^2$, how long would it take it to reach the sun's surface?

7.5 Consider a small mass m initially at rest a huge distance R from the sun. Assuming that θ, φ remain constant, show that the relevant Equations (7.14) and (7.16) from Section 7.5 become

$$-(1 - \gamma r^{-1})^{-1}\left(\frac{dr}{d\tau}\right)^2 + (1 - \gamma r^{-1})\left(\frac{dt}{d\tau}\right)^2 = 1$$

$$(1 - \gamma r^{-1})\frac{dt}{d\tau} = \beta.$$

(We have not made the text's assumption that $\gamma = 2GM$.) Conclude that

$$\left(\frac{dr}{dt}\right)^2 = (1 - \gamma r^{-1})^2 - \beta^{-2}(1 - \gamma r^{-1})^3.$$

Since we are assuming that initially $dr/dt = 0$, deduce that

$$f(r^{-1}) = \left(\frac{dr}{dt}\right)^2 = (1 - \gamma r^{-1})^2 - (1 - \gamma r_0^{-1})^{-1}(1 - \gamma r^{-1})^3,$$

with $f'(r_0^{-1}) = \gamma(1 - \gamma r_0^{-1})$. Of course, classically $\frac{1}{2}m(dr/dt)^2$, the kinetic energy, equals the loss of potential energy, $GMm(1/r - 1/r_0)$, so

$$f(r^{-1}) = \left(\frac{dr}{dt}\right)^2 = 2GM\left(\frac{1}{r} - \frac{1}{r_0}\right)$$

with $f'(r_0^{-1}) = 2GM$. Assuming the theories agree asymptotically for large r_0, conclude that $\gamma = 2GM$.

CHAPTER 8

The Gauss-Bonnet Theorem

One of the most celebrated results in mathematics, the Gauss-Bonnet Theorem, links the geometry and topology of surfaces. This chapter provides an overview without many proofs. For some fascinating physical interpretations and applications (see Levi [Lev1]).

8.1. The Gauss-Bonnet formula. *Let R be a smooth disc in a smooth two-dimensional Riemannian manifold M with Gauss curvature G. Let κ_g denote the geodesic curvature of the boundary. Then*

$$\int_R G + \int_{\partial R} \kappa_g = 2\pi. \tag{8.1}$$

For example, for a disc in the plane, $0 + 2\pi = 2\pi$. For the upper half of the unit sphere, $2\pi + 0 = 2\pi$.

Notice that this formula implies that G is intrinsic, as announced by Gauss's Theorema Egregium 3.6. The proof, like that of the Theorema Egregium, is a messy computation. It begins with a formula for G in local coordinates and changes $\int_R G$ into an integral over ∂R by Green's Theorem.

If ∂R has corners with interior angles α_i, as in Figure 8.1, then the boundary curvature term $\int_{\partial R} \kappa_g$ in Formula (8.1) may be reinterpreted to

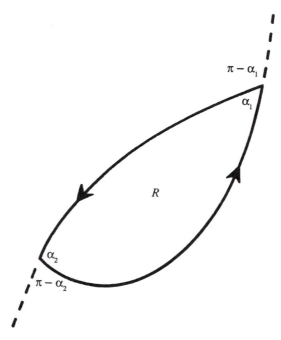

Figure 8.1. An interior angle α contributes $\pi - \alpha$ to $\int_{\partial R} \kappa_g$.

include the discrete contributions $\sum(\pi - \alpha_i)$. Alternatively, if the angles are treated separately,

$$\int_R G + \int_{\partial R} \kappa_g + \sum(\pi - \alpha_i) = 2\pi \qquad (8.2)$$

In particular, for a geodesic triangle $\triangle$,

$$\int_\triangle G + \pi = \alpha_1 + \alpha_2 + \alpha_3, \qquad (8.3)$$

a happy variation on the familiar statement that for a planar triangle the angles sum to π. By using triangles shrinking down to a point, we may compute the Gauss curvature as

$$G = \lim \frac{\alpha_1 + \alpha_2 + \alpha_3 - \pi}{\text{area } \triangle}.$$

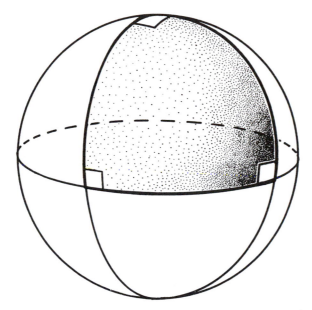

Figure 8.2. For a spherical triangle, the sum of the angles $\alpha_1 + \alpha_2 + \alpha_3 = \pi + A$. Here $\pi/2 + \pi/2 + \pi/2 = \pi + \pi/2$.

On the unit sphere, $\int_\triangle G$ becomes the area A of $\triangle$:

$$\alpha_1 + \alpha_2 + \alpha_3 = \pi + A, \tag{8.4}$$

the basic formula of spherical trigonometry. For example, for the geodesic triangle of Figure 8.2 with one vertex at the north pole, two on the equator, and three right angles,

$$\frac{\pi}{2} + \frac{\pi}{2} + \frac{\pi}{2} = \pi + \frac{\pi}{2}.$$

Gauss originally obtained Formula (8.3) in 1827. Bonnet provided Formula (8.1) in 1848.

8.2. The Gauss-Bonnet Theorem. The Gauss-Bonnet Theorem is a global result about a compact, two-dimensional smooth Riemannian manifold M. It relates a geometric quantity, the integral of the Gauss curvature,

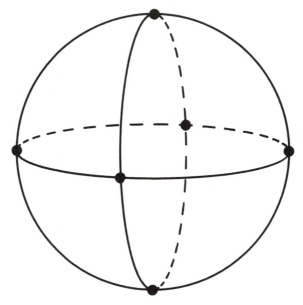

Figure 8.3. The unit sphere has Euler characteristic $\chi = V - E + F = 6 - 12 + 8 = 2$. The Gauss-Bonnet Theorem says that $\int G = 4\pi = 2\pi\chi$.

to a topological quantity, the Euler characteristic χ. For any triangulation of M, with V vertices, E edges, and F faces, χ is defined by $\chi = V - E + F$. The theorem says that

$$\int_M G = 2\pi\chi. \tag{8.5}$$

For example, consider the unit sphere, triangulated by the equator and two orthogonal great circles of longitude. (See Figure 8.3.) The Euler characteristic is

$$\chi = V - E + F = 6 - 12 + 8 = 2.$$

Hence

$$\int G = 4\pi = 2\pi\chi.$$

One remarkable consequence of Equation (8.5) is that the Euler characteristic is independent of the choice of triangulation and hence is a topological invariant. Actually, for a surface of genus g, the Euler characteristic $\chi = 2 - 2g$.

A second remarkable consequence of Equation (8.5) is that $\int G$ is independent of the metric, depending only on the topology of M.

Proof of Gauss-Bonnet Theorem. Fix a triangulation of M. On each triangle $\triangle$, the Gauss-Bonnet formula, Equation (8.2), becomes

$$\int_{\triangle} G = - \int_{\partial \triangle} \kappa_g + \sum \alpha_i - \pi.$$

Now add up the formulas for all the triangles. The first term contributes $\int_M G$. Since each edge occurs twice in opposite directions, the various $\int_{\partial \triangle} \kappa_g$ cancel. The angles around each vertex sum to 2π, so the angle term contributes $2\pi V$. The last term contributes πF. Finally, since each face has three edges and each edge lies on two faces, $E = \frac{3}{2}F$. Therefore

$$\int_M G = 2\pi V - \pi F = 2\pi\left(V - \frac{3}{2}F + F\right) = 2\pi(V - E + F) = 2\pi\chi.$$

8.3. The Gauss map of a surface in $\mathbf{R}^3$. The *Gauss map* of a surface M in $\mathbf{R}^3$ is just the unit normal $\mathbf{n} : M \to S^2$. Consider such a surface as pictured in Figure 8.4, tangent to the x, y-plane at the origin p_1, with principal curvatures κ_1 along the x-axis and κ_2 along the y-axis. For the purposes of illustration, suppose $\kappa_1 < 0$ and $\kappa_2 > 0$.

We want to consider the derivative $D\mathbf{n}$, called the *Weingarten map*. If we move in the x-direction from p_1 toward a point p_2, $\mathbf{n}$ turns in the x-direction an amount proportional to $|\kappa_1|$, but this amount is positive, whereas κ_1 is negative. Indeed, the first column of $D\mathbf{n}$ is $\begin{bmatrix} -\kappa_1 \\ 0 \end{bmatrix}$. If we move instead in the y-direction from p_1 toward a point p_3, $\mathbf{n}$ turns in the negative y-direction an amount proportional to $|\kappa_2|$. The second column of $D\mathbf{n}$ is $\begin{bmatrix} 0 \\ -\kappa_2 \end{bmatrix}$. Hence

$$D\mathbf{n} = \begin{bmatrix} -\kappa_1 & 0 \\ 0 & -\kappa_2 \end{bmatrix} = -\mathrm{II}.$$

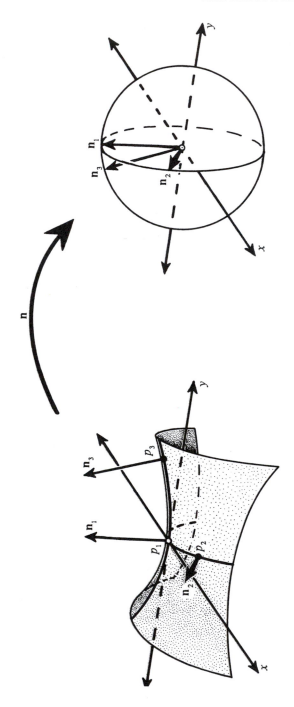

Figure 8.4. The Gauss map sends a point to the unit normal in the sphere.

This identity holds in any orthonormal coordinates. The Jacobian of the Gauss map equals the Gauss curvature:

$$\det D\mathbf{n} = \kappa_1 \kappa_2 = G.$$

If M is a topological sphere, $\mathbf{n}$ has degree 1 (that is, covers the sphere once, algebraically), and

$$\int_M G = \text{area (image } \mathbf{n}) = 4\pi = 2\pi\chi.$$

We have recovered the Gauss-Bonnet Theorem for a sphere in $\mathbf{R}^3$. For general compact M in $\mathbf{R}^3$, $\mathbf{n}$ has degree $\chi/2$, and

$$\int_M G = \frac{\chi}{2}4\pi = 2\pi\chi,$$

the Gauss-Bonnet Theorem for any closed surface in $\mathbf{R}^3$.

8.4. The Gauss map of a hypersurface. For a hypersurface M^n in $\mathbf{R}^{n+1}$, the Gauss map $\mathbf{n} : M \to S^n$. In orthonormal coordinates aligned with the principal curvature directions at a point, the Weingarten map is

$$D\mathbf{n} = \begin{bmatrix} -\kappa_1 & & 0 \\ & \ddots & \\ 0 & & -\kappa_n \end{bmatrix} = -\text{II},$$

and the Jacobian of $\mathbf{n}$ is

$$(-\kappa_1)\cdots(-\kappa_n) = (-1)^n G,$$

if the Gauss curvature G is defined as $\kappa_1 \cdots \kappa_n = \det \text{II}$. As for surfaces, if n is even, the degree of $\mathbf{n}$ is the Euler characteristic

$$\chi = V - E + F - \cdots$$

and

$$\int_M G = \frac{\chi}{2} \text{ area } S^n, \tag{8.6}$$

a generalization of Gauss-Bonnet to hypersurfaces by H. Hopf in 1925 [Hop]. (If n is odd, $\chi = 0$.)

8.5. The Gauss-Bonnet-Chern Theorem. Amazingly enough, a generalization of the Gauss-Bonnet Theorem (8.5) holds for any even-dimensional smooth compact Riemannian manifold M. An extrinsic proof was obtained by C. B. Allendoerfer [All] and W. Fenchel [Fen] around 1938, an intrinsic proof by S.-S. Chern [Che2] in 1944. (See also Gottlief [Got] and Spivak [Spi, Volumes III, V].)

The formulation and proof require a definition of G in local coordinates. It is

$$G = \frac{1}{2^{n/2}n!\,\det g_{ij}} R_{i_1 i_2 j_1 j_2} R_{i_3 i_4 j_3 j_4} \cdots R_{i_{n-1} i_n j_{n-1} j_n} \epsilon^{i_1 \cdots i_n} \epsilon^{j_1 \cdots j_n},$$

where $\epsilon^{i_1 \cdots i_n} = \pm 1$, according to whether $i_1, \cdots, i_n$ is an even or odd permutation. For example, for a two-dimensional surface tangent to the x_1, x_2-plane at 0 in $\mathbf{R}^n$, with x_1, x_2 as local coordinates, $\det g_{ij} = 1$ and

$$G = \frac{1}{2^1 2!}(R_{1212} - R_{1221} - R_{2112} + R_{2121}) = R_{1212},$$

the Gauss curvature of the only section there is (compare to Equation (5.3)).

Actually Chern used the language of differential forms and moving frames. He defined G as the Pfaffian (a square root of the determinant) of certain curvature forms. His pioneering work on fiber bundles launched the modern era in differential geometry.

8.6. Parallel transport. A vectorfield on a curve is called *parallel* if its covariant derivative along the curve vanishes (see Equation (6.16)). A vector at a point on a curve can be uniquely continued "by parallel transport" as a parallel vectorfield. In Euclidean space, a parallel vectorfield is constant— that is, the vectors are all "parallel."

In a Riemannian manifold M, a curve is a geodesic if and only if its unit tangent T is parallel. If M is a two-dimensional surface, γ is a curve, and θ is the angle from a parallel vectorfield X to the unit tangent T,

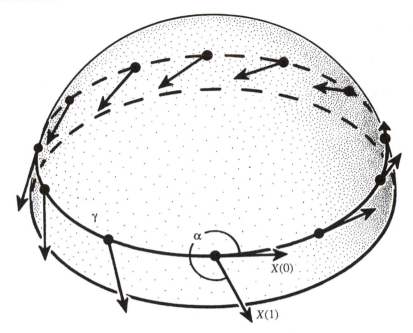

Figure 8.5. Geodesic curvature $\kappa_g = d\theta/ds$, where θ is the angle from a parallel vectorfield X to the unit tangent $\mathbf{T}$. By the Gauss-Bonnet formula, the angle α from the initial $X(0)$ to the final $X(1)$ equals $\int G$. For example, heading east along a circle of latitude in the northern unit hemisphere involves curving to the left (think of a small circle around the north pole). For latitude near the equator, this effect is small, and a *parallel* vectorfield ends up pointing slightly to the right, i.e., at an angle α of almost 2π to the left. Sure enough, the enclosed area also is almost 2π, the area of the whole northern hemisphere.

then the geodesic curvature $\kappa_g = d\theta/ds$. If γ is a closed curve, the result $X(1)$ of parallel-transporting X around the curve will be at some angle α from the starting vector $X(0)$. (See Figure 8.5.) By the Gauss-Bonnet Formula (8.1),

$$2\pi - \int G = \int \kappa_g = \int \frac{d\theta}{ds} = 2\pi - \alpha,$$

so $\alpha = \int G$. Hence the Gaussian curvature may be interpreted as the net amount a vector turns under parallel transport around a small closed curve.

More generally, in a higher-dimensional Riemannian manifold M, R_{ijkl} may be interpreted as the amount a vector turns in the e_i, e_j-plane under parallel transport around a small closed curve in the e_k, e_l-plane.

We have already seen the infinitesimal version of this interpretation of Riemannian curvature in Formula (6.13):

$$X^i_{;k;l} - X^i_{;l;k} = -\sum_j R^i_{jkl} X^j.$$

The left-hand side describes the effects on X of moving in an infinitesimal parallelogram: first in the k direction, then in the l direction, then in the $-k$ direction, then in the $-l$ direction. R^i_{jkl} gives the amount the j component of the original vector X contributes to the i component of the change in X.

8.7. A proof of Gauss-Bonnet in $\mathbf{R}^3$.

Ambar Sengupta has shown me a simple proof of the Gauss-Bonnet Formula (8.1), for surfaces in $\mathbf{R}^3$. Then, of course, the Gauss-Bonnet Theorem 8.5 follows easily as in Section 8.2.

The proof begins with a simple proof of the formula for a geodesic triangle on the unit sphere Equation (8.4),

$$\alpha_1 + \alpha_2 + \alpha_3 = \pi + A, \tag{8.7}$$

due to Thomas Harriot (1603, see Lohne [Loh, p.301]). The formula for a smooth disc-type region R on the sphere follows by approximation:

$$\text{area}(R) + \int_{\partial R} \kappa_g = 2\pi. \tag{8.8}$$

Finally, an ingenious argument deduces the formula for a smooth disc-type region on any smooth surface M in $\mathbf{R}^3$:

$$\int_R G + \int_{\partial R} \kappa_g = 2\pi. \tag{8.9}$$

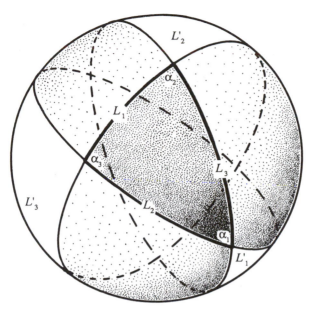

Figure 8.6. On a unit sphere, the sum $\alpha_1 + \alpha_2 + \alpha_3$ of the angles of a geodesic triangle equals $\pi + A$, as may be proved by viewing the triangle as the intersection of three lunes L_i, each of area $2\alpha_i$.

To prove Equation (8.7), consider a geodesic triangle $\triangle$ of area A and angles $\alpha_1, \alpha_2, \alpha_3$, bounded by three great circles as in Figure 8.6. Each pair of great circles bounds two congruent lunes L_i, L_i' with angles α_i. The lunes L_i intersect in $\triangle$; the lunes L_i' intersect in a congruent triangle $\triangle'$ on the back of the sphere. The lune L_i has area proportional to α_i; consideration of the extreme case $\alpha_i = \pi$ shows that $\text{area}(L_i) = 2\alpha_i$. Since $\cup L_i$ is congruent to $\cup L_i'$ each has area 2π. Hence

$$2\pi = \text{area}(\cup L_i) = \sum \text{area}(L_i) - 2A = 2(\alpha_1 + \alpha_2 + \alpha_3) - 2A.$$

Therefore $\alpha_1 + \alpha_2 + \alpha_3 = \pi + A$, as desired.

From piecing together geodesic triangles it follows that for any geodesic polygon on the sphere,

$$\text{area}(R) + \sum (\pi - \alpha_i) = 2\pi.$$

To deduce Equation (8.9), consider the Gauss map $\mathbf{n} : M \to \mathbf{S}^2$. Since the Jacobian is the Gauss curvature G, the Gauss map $\mathbf{n}$ maps the region R to a region R' of area $A = \int_R G$.

Let $\gamma(t)$ $(0 \le t \le 1)$ be the curve bounding R. Let $X(t)$ be a parallel vectorfield on γ, so $\dot{X}(t)$ is a multiple of $\mathbf{n}$. Along the curve $\gamma'(t) = \mathbf{n} \circ \gamma(t)$ on the sphere, the unit normal is the same, so $X(t)$, bodily moved in $\mathbf{R}^3$ to the sphere, is still parallel.

Let α be the angle from $X(0)$ to $X(1)$. Then $\int_{\partial R} \kappa_g - 2\pi$ and $\int_{\partial R'} \kappa_g - 2\pi$ both equal $-\alpha$ (see Figure 8.5). Therefore

$$\int_R G + \int_{\partial R} \kappa_g - 2\pi = \text{area}(R') + \int_{\partial R'} \kappa_g - 2\pi = 0$$

by Equation (8.8), proving Equation (8.9).

Here area (R') denotes the algebraic area of R', negative if G is negative and $\mathbf{n}$ reverses orientation. Similarly, $\int_{\partial R'} \kappa_g - 2\pi$ must be interpreted so that, for example, if G is negative, it switches sign too.

EXERCISES

8.1. Try to check directly Gauss's formula for a geodesic triangle $\alpha_1 + \alpha_2 + \alpha_3 = \pi + \int G$ for any isosceles geodesic triangle on the sphere. (You might put one vertex at the north pole.)

8.2. Compute that the Euler characteristic of the sphere is 2, of the torus is 0, and of the two-holed torus is -2 by using triangulations.

8.3. Try to prove directly that the Euler characteristic is independent of triangulation.

8.4. Consider the round two-dimensional sphere S in $\mathbf{R}^3$ of radius a.

a. Use $dA = |\mathbf{x}_\varphi \wedge \mathbf{x}_\theta| d\varphi \, d\theta$ to show $\int_S G dA = (1/a^2) \int_S dA = 4\pi$.

b. Use $dA = \sqrt{\det g_{ij}} d\theta dz$ to get the same result.

8.5. Use $dA = \sqrt{\det g_{ij}} d\theta d\varphi$ to compute $\int_T G\, dA = 0$ for the torus T of Exercise 6.1.

8.6. One night at about 9:00 pm at Boston's Museum of Science I observed a pendulum swinging in a direction v, rotated $120°$ from the starting direction marked 9:00 am.

a. What is Boston's latitude?

b. Was v rotating clockwise or counterclockwise?

8.7. TRUE/FALSE

a. Every unit vector on the unit round sphere can be gotten from any other by parallel displacement along some curve.

b. Every unit vector on the unit round cylinder can be gotten from any other by parallel displacement along some curve.

8.8. *Review problem on the sphere.* Consider the round two-dimensional sphere $\mathbf{S}^2 \subset \mathbf{R}^3$ of radius a.

a. Find the sectional curvature, Ricci curvature, scalar curvature, scalar mean curvature, and mean curvature vector.

b. Compute the geodesic curvature of the equator.

c. Show that a circle of latitude for fixed $0 < \varphi < \pi/2$ has geodesic curvature $\kappa = \cos\varphi/(a\sin\varphi)$.

d. Check the Gauss-Bonnet formula for the enclosed polar region from **c.**

8.9. Consider the cone $C \subset \mathbf{R}^3$ defined by

$$z = \sqrt{x^2 + y^2}, \qquad 0 \le z \le 2.$$

a. Compute the mean and Gauss curvature at a general point. Then check your answer for the mean curvature using Exercise 3.7.

b. What are the Ricci and scalar curvature R at a general point? What is the Einstein tensor?

c. Compute the scalar geodesic curvature κ_g of a circle in C at height z.

d. Compute the area of the part of C inside a ball about 0 of radius $r \leq 2$.

e. The part of C with $0 \leq z \leq 1$ is a topological disc. Compute that the Gauss-Bonnet formula is false for this disc. What hypothesis of the Gauss-Bonnet formula does not hold?

f. Find the length of the shortest geodesic in C from the point $(0, -1, 1)$ to the point $(0, 1, 1)$.

CHAPTER 9

Geodesics and Global Geometry

Our streamlined approach has avoided a deep study of geodesics or even the exponential map. This chapter discusses geodesics and some theorems that draw global conclusions from local curvature hypotheses. For example, Bonnet's Theorem 9.5 obtains a bound on the diameter of M from a bound on the sectional curvature. Cheeger and Ebin [Che1] provide a beautiful reference on such topics in global Riemannian geometry.

Let M be a smooth Riemannian manifold. Recall that by the theory of differential equations, there is a unique geodesic through every point in every direction. Assume that M is (geodesically) *complete*—that is, geodesics may be continued indefinitely. (The geodesic may overlap itself, as the equator winds repeatedly around the sphere.) This condition means that M has no boundary and no missing points.

9.1. The exponential map. The *exponential map* Exp_p at a point p in M maps the tangent space T_pM into M by sending a vector $\mathbf{v}$ in T_pM to the point in M a distance $|\mathbf{v}|$ along the geodesic from p in the direction $\mathbf{v}$. (See Figure 9.1.) For example, let M be the unit circle in the complex plane $\mathbf{C}$, $p = 1, T_pM = \{iy\}$. Then

$$\mathrm{Exp}_1(iy) = e^{iy}.$$

(See Figure 9.2.)

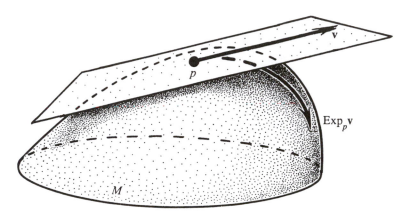

Figure 9.1. The exponential map Exp_p maps $\mathbf{v} \in T_pM$ to
the point a distance $|\mathbf{v}|$ along the geodesic in the direction $\mathbf{v}$.

As a second example, let M be the Lie group $SO(n)$ of rotations of
$\mathbf{R}^n$, represented as

$$SO(n) = \{n \times n \text{ matrices } A : AA^t = I \text{ and } \det A = 1\} \subset \mathbf{R}^{n^2}.$$

The tangent space at the identity matrix I consists of all skew-symmetric
matrices,

$$T_I SO(n) = \{A : A^t = -A\},$$

because differentiating the defining relation $AA^t = I$ yields $IA^t + AI^t =$
0, that is, $A^t = -A$. (See Figure 9.3.)

The exponential map on $T_I SO(n)$ is given by the exponential matrix
function familiar from linear algebra:

$$\text{Exp}_I(A) = e^A = I + A + \frac{A^2}{2!} + \cdots .$$

For any point p in a smooth Riemannian manifold M, $\exp_p$ is a smooth
diffeomorphism at 0. It provides very nice coordinates called *normal
coordinates* in a neighborhood of p. Normal coordinates have the useful
property that the metric $g_{ij} = I$ to first order at p. (Compare to Theorem
3.6.)

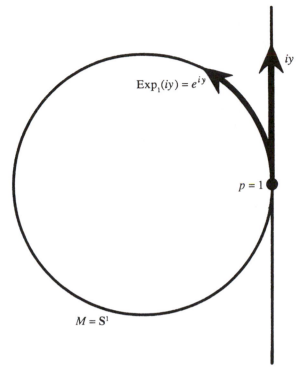

Figure 9.2. For M the circle, the tangent space is $T_1 M = \{iy\}$, and the exponential map is $\text{Exp}_1(iy) = e^{iy}$.

A small open ball in normal coordinates is simple and convex: there is a unique geodesic between any two points. (*Simple* means at most one; *convex* means at least one.) Moreover, that geodesic is the shortest path in all of M between the two points.

The Hopf-Rinow Theorem says that as long as M is connected, there is a geodesic giving the shortest path between any two points. In particular, Exp_p maps $T_p M$ onto M.

9.2. The curvature of SO(n). As an example, we now compute the curvature of $SO(n)$. In Chapter 5 we defined the second fundamental tensor of a submanifold M of $\mathbf{R}^n$ by the turning rate κ of unit tangents along each slice curve. As long as we take the normal component, any

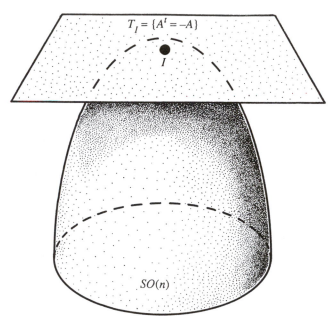

Figure 9.3. For the special orthogonal group $SO(n)$, the tangent space is $T_I SO(n) = \{A : A^t = -A\}$, and the exponential map is $\mathrm{Exp}_I(A) = e^A$.

curve heading in the same direction and any tangent vectorfield starting with the same tangent vector will give the same result.

Let $\{E_i\}$ be an orthonormal basis for $T_I SO(n) = \{A = -A^t\}$. The ij component of II may be computed as the normal component of the derivative along the curve e^{sE_i} of the vectorfield $e^{sE_i}E_j$. (Since e^{sE_i} maps I to e^{sE_i}, it maps E_j to a tangent vector $(e^{sE_i})E_j$.) First we compute

$$\frac{d}{ds}(e^{sE_i}E_j) = \frac{d}{ds}(1 + sE_i + \cdots)E_j = E_i E_j.$$

The projection onto the normal space of symmetric matrices is given by

$$\frac{1}{2}[(E_i E_j) + (E_i E_j)^t] = \frac{1}{2}(E_i E_j + E_j E_i).$$

Hence

$$\mathrm{II} = \frac{1}{2}(E_i E_j + E_j E_i).$$

The curvature of the E_i, E_j section is given by

$$K(E_i, E_j) = \frac{1}{4}[2E_i^2 \cdot 2E_j^2 - (E_i E_j + E_j E_i) \cdot (E_i E_j + E_j E_i)].$$

Since for matrices $A \cdot B = \mathrm{trace}\,(AB^t)$ and $(AB)^t = B^t A^t$,

$$
\begin{aligned}
K(E_i, E_j) &= \frac{1}{4}\mathrm{trace}\,(4E_i E_i E_j^t E_j^t - E_i E_j E_j^t E_i^t \\
&\quad - E_i E_j E_i^t E_j^t - E_j E_i E_j^t E_i^t - E_j E_i E_i^t E_j^t) \\
&= \frac{1}{4}\mathrm{trace}\,(4E_i E_i E_j E_j - E_i E_i E_j E_j \\
&\quad - E_i E_j E_i E_j - E_i E_j E_i E_j - E_i E_i E_j E_j),
\end{aligned}
$$

because $E_i^t = -E_i$ and $\mathrm{trace}\,(AB) = \mathrm{trace}\,(BA)$, (although trace $(ABC) \neq \mathrm{trace}\,(CBA)$). Hence

$$
\begin{aligned}
K(E_i, E_j) &= \frac{1}{4}\mathrm{trace}\,(E_i E_j E_j E_i - E_i E_j E_i E_j \\
&\quad - E_j E_i E_j E_i + E_j E_i E_i E_j) \\
&= \frac{1}{4}\mathrm{trace}\,([E_i, E_j][E_i, E_j]^t),
\end{aligned}
$$

where $[E_i, E_j]$ denotes the bracket product $E_i E_j - E_j E_i$. Therefore

$$K(E_i, E_j) = \frac{1}{4}\|[E_i, E_j]\|^2.$$

Indeed in any compact Lie group, for orthonormal vectors $\mathbf{v}, \mathbf{w}$,

$$K(\mathbf{v}, \mathbf{w}) = \frac{1}{4}\|[\mathbf{v}, \mathbf{w}]\|^2$$

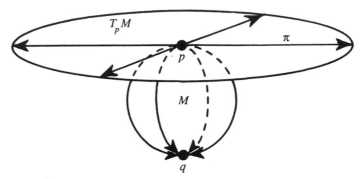

Figure 9.4. On the sphere M, Exp_p maps an open disc diffeo-
morphically onto $M - \{q\}$, but maps the whole boundary circle
onto $\{q\}$. The singular point q for $\exp_p$ is called a conjugate
point.

(see Milnor [Mil, Line 7.3, p. 323], or Helgason [Hel, Exercises II.A.6(iv)
and IV.A.2(ii)]). For example, for $SO(3)$, take

$$E_1 = \frac{1}{\sqrt{2}} \begin{bmatrix} 0 & -1 & 0 \\ 1 & 0 & 0 \\ 0 & 0 & 0 \end{bmatrix},$$

$$E_2 = \frac{1}{\sqrt{2}} \begin{bmatrix} 0 & 0 & -1 \\ 0 & 0 & 0 \\ 1 & 0 & 0 \end{bmatrix},$$

$$E_3 = \frac{1}{\sqrt{2}} \begin{bmatrix} 0 & 0 & 0 \\ 0 & 0 & -1 \\ 0 & 1 & 0 \end{bmatrix}.$$

Then

$$K(E_1, E_2) = \frac{1}{4} \left| [E_1, E_2] \right|^2$$

$$= \frac{1}{4} \left| \begin{bmatrix} 0 & 0 & 0 \\ 0 & 0 & -\frac{1}{2} \\ 0 & \frac{1}{2} & 0 \end{bmatrix} \right|^2 = \frac{1}{8}.$$

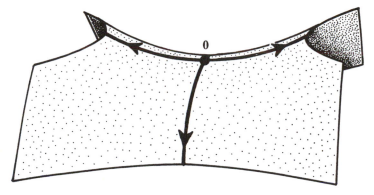

Figure 9.5. For the saddle $\{z = -x^2 + y^2\}$, Exp_0 is a global diffeomorphism.

Indeed, all the sectional curvatures turn out to be $\frac{1}{8}$. Actually $SO(3) =$ $\mathbf{R}P^3$ is just a round 3-sphere of radius $2\sqrt{2}$ with antipodal points identified.

For $SO(n), 0 \le K \le \frac{1}{8}$.

9.3. Conjugate points and Jacobi fields. Although Exp_p is a diffeomorphism at 0, it need not be a diffeomorphism at all points $\mathbf{v} \in T_p M$. For example, let M be the unit sphere and let p be the north pole. Then Exp_p maps the disc $\{\mathbf{v} \in T_p M : |\mathbf{v}| < \pi\}$ diffeomorphically onto $M - \{q\}$, where q is the south pole, but it maps the whole circle $\{|\mathbf{v}| = \pi\}$ onto $\{q\}$. (See Figure 9.4.)

On the other hand, for the saddle $\{z = -x^2 + y^2\}$ of Figure 9.5, Exp_0 is a global diffeomorphism.

A point $q = \text{Exp}_p \mathbf{v} \in M$ is called *conjugate* to p if Exp_p fails to be a diffeomorphism at $\mathbf{v}$—that is, if the linear map $D\text{Exp}_p \mathbf{v}$ is singular. This occurs when moving perpendicular to $\mathbf{v}$ at $\mathbf{v} \in T_p M$ corresponds to zero velocity at $q \in M$, or roughly when nearby geodesics from p focus at q. Such conjugate points q are characterized by a variation "Jacobi" vectorfield J along the geodesic, vanishing at p and q, which provides the initial velocity for finding nearby geodesics (see Spivak [Spi, Volume 4, Chapter 8]).

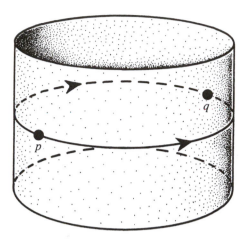

Figure 9.6. Geodesics on the cylinder originating at p stop being shortest paths at the cut point q.

Note. It turns out that once a geodesic passes a conjugate point, it is no longer the shortest geodesic from p.

We state the following theorem as an early example of the relationship between curvature and conjugate points (see Cheeger and Ebin [Che1, Rauch's Theorem 1.28]).

Theorem. *Let M be a smooth Riemannian manifold. If the sectional curvature K at every point for every section is bounded above by a constant K_0, then the distance from any point to a conjugate point is at least $\pi/\sqrt{K_0}$. In particular, if the sectional curvature K is nonpositive, there are no conjugate points, and Exp_p is a (local) diffeomorphism at every point.* (We say that Exp_p is a *submersion* or a *covering map*.)

9.4. Cut points and injectivity radius. A *cut point* is the last point on a geodesic from p to which the geodesic remains the shortest path from p. The cut point q could be conjugate to p, as the antipodal point on a sphere, where infinitesimally close geodesics focus (see the Note above). Alternatively, the cut point q could be like the antipodal point on

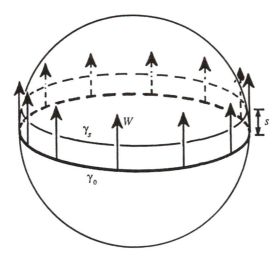

Figure 9.7. The second variation of the length of the equator
$L''(0) = -\int K(T, W) = -2\pi$.

the cylinder of Figure 9.6, where geodesics heading in opposite directions from p eventually meet.

Inside the locus of cut points, Exp_p is injective, a diffeomorphism. The infimum of distances from any point to a cut point is called the *injectivity radius* of the manifold. For example, the injectivity radius of a cylinder of radius a is πa.

Bounding the sectional curvature does not bound the injectivity radius away from 0. A cylinder with Gauss curvature 0 can have an arbitrarily small radius and injectivity radius. Likewise, hyperbolic manifolds with negative curvature can have a small injectivity radius. A common hypothesis for global theorems is *bounded geometry*: sectional curvature bounded above and injectivity radius bounded below.

9.5. Bonnet's Theorem. Bonnet's Theorem draws a global conclusion from a local, curvature hypothesis:

Let M be a smooth (connected) Riemannian manifold with sectional curvature bounded below by a positive constant K_0. Then the diameter of M is at most $\pi/\sqrt{K_0}$.

The *diameter* of M is the greatest distance between any two points. The unit sphere with $K = 1$ and diameter π (distance is measured on the sphere) shows that Bonnet's Theorem is sharp.

We now give a proof sketch beginning with three lemmas. The first lemma relates the second variation of the length of a geodesic to sectional curvature K.

Lemma. *Let γ be a finite piece of geodesic with unit tangent T. Let W be an orthogonal, parallel unit variational vectorfield on γ. Then the initial second variation of length is given by*

$$L''(0) = - \int K(T, W). \tag{9.1}$$

For example, let γ be the equator on the unit sphere of Figure 9.7. Let W be the unit upward vectorfield. Then the circle of latitude γ_s a distance s from γ has length $L(s) = 2\pi \cos s$. Hence

$$L''(0) = -2\pi = - \int K,$$

since $K = 1$.

This lemma illustrates our earlier remark (Section 6.7) that positive curvature means that parallel geodesics converge (and hence cross-sectional distance decreases).

Note that by scaling, for the variational vectorfield aW of length a,

$$L''(0) = \int -a^2 K(T, W). \tag{9.2}$$

The second lemma considers variational vectorfields of variable length.

Lemma. *Let γ be an initial segment of the x-axis in $\mathbf{R}^2$ of length $L(0)$. Consider a smooth vertical variational vectorfield $f(x)\mathbf{j}$. Then the initial second variation of length is given by*

$$L''(0) = \int\limits_0^{L(0)} f'(x)^2 dx. \tag{9.3}$$

Proof. Flowing $sf(x)\mathbf{j}$ produces a curve of length

$$L(s) = \int_0^{L(0)} [1 + s^2 f'(x)^2]^{1/2} dx$$

$$= \int_0^{L(0)} [1 + s^2 f'(x)^2 + \cdots] dx.$$

Differentiation yields Lemma (9.3).

The third lemma states without further proof the result of combining the effects of the first two Lemmas (9.2), (9.3).

Lemma. *Let γ be a finite piece of geodesic with unit tangent T. Consider a variational vectorfield fW, where W is an orthogonal, parallel unit variational vectorfield on γ. Then the initial second variation of length is given by*

$$L''(0) = \int_\gamma [f'^2 - f^2 K(T, W)]. \tag{9.4}$$

Proof of Bonnet's Theorem. Suppose diam $M > \pi/\sqrt{K_0}$. Then there is some shortest geodesic $\gamma(t)$ of length $l > \pi/\sqrt{K_0}$. Hence $K \geq K_0 > \pi^2/l^2$.

Assume γ is parameterized by arc length t. Let W be an orthogonal, parallel unit vectorfield on γ. Take as a variational vectorfield $(\sin \frac{\pi}{l} t) W$, which vanishes at the endpoints of γ. By Equation (9.4), the initial second variation of length is given by

$$L''(0) = \int_0^l \left(\frac{\pi}{l} \cos \frac{\pi}{l} t\right)^2 - \left(\sin^2 \frac{\pi}{l} t\right) K(T, W)$$

$$< \int_0^l \frac{\pi^2}{l^2} \cos^2 \frac{\pi}{l} t - \frac{\pi^2}{l^2} \sin^2 \frac{\pi}{l} t = 0.$$

This contradiction of the choice of γ as a *shortest* path completes the proof.

Remark. In the proof of Bonnet's Theorem, we could have chosen any unit vector orthogonal to T for W at the starting point of γ (extending W by parallel transport). Averaging over all such choices permits us to replace the bound on K by a bound on its average, the Ricci curvature. The theorem of Myers concludes that if Ric $\geq (n-1)K_0$, then diam $M \leq \pi/\sqrt{K_0}$.

9.6. Constant curvature, the Sphere Theorem, and the Rauch Comparison Theorem.
This section just mentions some famous results on a smooth, connected, complete Riemannian manifold M.

Suppose that M is simply connected, so every loop can be shrunk to a point. Suppose the sectional curvature K is constant for all sections at all points. By scaling, we may assume K is 1, 0, or -1.

If $K = 1, M$ is the unit sphere. If $K = 0, M$ is Euclidean space. If $K = 1, M$ is hyperbolic space. Thus the metric and global geometry are completely determined for constant curvature.

The Sphere Theorem, perhaps the most famous global theorem, draws topological conclusions from hypotheses that the curvature is "pinched" between two values.

The Sphere Theorem. *Let M be simply connected with sectional curvature $\frac{1}{4} < K \leq 1$. Then M is a topological sphere (homeomorphic to the standard sphere).*

The theorem is sharp, since, for example, complex projective space CP^2 has $\frac{1}{4} \leq K \leq 1$. It was proved by H. Rauch for $\frac{3}{4} \leq K \leq 1$ in 1951, and generalized to $\frac{1}{4} \leq K \leq 1$ by M. Berger and W. Klingenberg in 1960.

The Rauch Comparison Theorem. One of the main ingredients in a proof, and one of the most useful tools in Riemannian geometry, is the Rauch Comparison Theorem. It says, for example, the following:

Let M_1, M_2 be complete smooth Riemannian manifolds with sectional curvatures $K_1 \geq K_0 \geq K_2$ for some constant K_0. For $p_1 \in M_1$, $p_2 \in M_2$, identify $T = T_{p_1} M_1 = T_{p_2} M_2$ via a linear isometry. Let B be an open ball about 0 in T on which Exp_{p_1} and Exp_{p_2} are diffeomorphisms into M_1 and M_2. Let γ be a curve in B, and let γ_1, γ_2 be its images in M_1, M_2. Then length $(\gamma_1) \leq$ length (γ_2).

In applications, either M_1 or M_2 is usually taken to be a sphere, Euclidean space, or hyperbolic space, all of which have well-known trigonometries. Thus one obtains distance estimates on the other manifold from curvature bounds.

An important and stronger comparison theorem is due to Toponogov.

9.7. The isoperimetric problem. One standard isoperimetric problem seeks the shortest simple closed curve enclosing a region of prescribed area in a Riemannian surface. (The name comes from the often equivalent problem of seeking the largest area enclosed by curves of equal perimeter.) The interesting survey by Howards, Hutchings, and Morgan [How, Introduction, Section 8] describes the following new theorem after Benjamini and Cao [Ben, Theorems 5, 6], who give the first proof that the isoperimetric solution in the paraboloid of revolution

$$P = \{z = x^2 + y^2\} \subset \mathbf{R}^3$$

is a horizontal circle.

The existence of a nice shortest curve is an interesting question. In a compact surface, perhaps with boundary, the compactness properties of Lipschitz functions immediately produce a minimizer. A recent, simple argument of Hass and Morgan [Has2] shows that away from the boundary *the minimizer is an embedded curve of constant geodesic curvature κ_0, except possibly for finitely many geodesic arcs or isolated points where it bumps up against itself but remains C^1*. If one allows many components enclosing disjoint regions, the geodesic curvature $\kappa \leq \kappa_0$ everywhere.

Given such existence and regularity, one can prove quite easily the following isoperimetric theorem.

Theorem ([How, Section 8], [Pan, Proposition 7]). *Consider the plane with smooth, rotationally symmetric, complete metric such that the Gauss curvature is a strictly decreasing function of the distance from the origin. Then the unique length-minimizing simple closed curve enclosing a given area is a circle centered at the origin.*

Lemma. *Let L(t) denote the least perimeter of $m \leq m_0$ components enclosing m disjoint regions of total area t. Then for $0 < t_1 \leq t \leq t_2$, there is a constant $a(t_1, t_2) > 0$ such that*

$$F(t) = L(t) + at \quad \text{is a nondecreasing function of } t \qquad (9.5)$$

(and in particular L(t) is differentiable almost everywhere).

Proof of lemma. Choose a such that $aL(t)$ is larger than the Gauss curvature of a region of area t. If $F(t)$ ever decreases, choose $s < t$ with $F(s) > F(t)$. Of course F is continuous. Let u be the smallest number larger than s with $F(u) = F(t)$. Then for $s < v < u$, $F(v) > F(u)$.

Consider a minimal enclosure of area u. By the Gauss-Bonnet Formula (8.1), $2\pi - \kappa_0 L(u)$ is less than or equal to the total Gauss curvature enclosed, which by choice of a is less than $aL(u)$. So $\kappa_0 > -a$, and we can remove a little area $u - v$ from the enclosure and increase its length by less than $a(u - v)$, which is a contradiction.

Proof of theorem. Inside a large ball, for m_0 large, let $L(t)$ denote the length of a shortest curve of $m \leq m_0$ components enclosing m disjoint regions of total area t. If L is differentiable at t, then $L'(t)$ is the geodesic curvature κ_0. Gauss-Bonnet tells us that the total Gauss curvature of the enclosed region equals

$$2\pi\chi - \int \kappa \geq 2\pi - L(t)\kappa_0 = 2\pi - L(t)L'(t).$$

Let $f(t)$ denote the total Gauss curvature of a disc of area t centered at the origin. Since the Gauss curvature is a decreasing function of radius, any other region with the same area must have less total Gauss curvature. So we have

$$2\pi - L(t)L'(t) \leq f(t),$$

$$L(t)L'(t) \geq 2\pi - f(t).$$

By Lemma (9.5), integration from $t = 0$ to T yields

$$L(T)^2 \geq 2 \int L(t)L'(t) \geq 4\pi T - 2 \int f(t).$$

This inequality is sharp for a circle centered at the origin (as we can see by integrating the Gauss-Bonnet formula for circles centered at the origin with area t from 0 to T) and strict otherwise.

CHAPTER 10
General Norms

In nature, the energy of a path or surface often depends on direction as well as length or area. The surface energy of a crystal, for example, depends radically on direction. Indeed, some directions are so much cheaper that most crystals use only a few cheap directions. (See Figure 10.1.) This chapter applies more general costs or norms Φ to curves and presents an appropriate generalization of curvature.

10.1. Norms. A norm Φ on $\mathbf{R}^n$ is a convex homogeneous function on $\mathbf{R}^n$, positive except that $\Phi(0) = 0$. We call Φ C^k if its restriction to $\mathbf{R}^n - \{0\}$ is C^k (or, equivalently, if its restriction to the unit sphere S^{n-1} is C^k). The convexity of Φ is equivalent to the convexity of its unit ball

$$\{x : \Phi(x) \le 1\}.$$

For any curve C, parametrized by a differentiable map $\gamma : [0, 1] \to \mathbf{R}^n$ define

$$\Phi(C) = \int_C \Phi(\mathbf{T})ds = \int_{[0,1]} \Phi(\dot{\gamma})dt.$$

If C is a straight line segment, then

$$\Phi(C) = \Phi(\mathbf{T}) \text{ length } C.$$

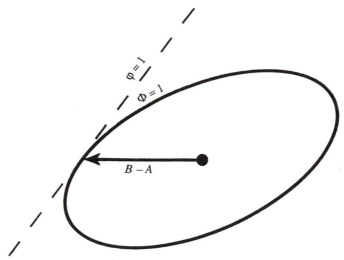

Figure 10.2. Since the unit ball of Φ is strictly convex, there is a linear function or 1-form φ such that $\varphi(v) \leq \Phi(v)$ with equality only if $v = B - A$.

10.2. Proposition. *Among all differentiable curves C from A to B, the straight line L minimizes $\Phi(C)$, uniquely if Φ is strictly convex.*

Proof. Since the unit ball of Φ is convex, there is a constant-coefficient differential form φ such that

$$\varphi(v) \leq \Phi(v),$$

with equality when $v = B - A$. (See Figure 10.2.) If Φ is strictly convex, equality holds only if v is a multiple of $B - A$. Let C' be any differentiable curve from A to B.

Figure 10.1. Crystal shapes typically have finitely many flat facets corresponding to surface orientation of low energy. (The first two photographs are from Steve Smale's *Beautiful Crystals Calendar*. The third photograph is from E. Brieskorn. All three appear in *The Parsimonious Universe* by S. Hildebrandt and A. Tromba ([Hil, pp. 263–264]).

Then

$$\Phi(C') = \int_{C'} \Phi(\mathbf{T})ds \geq \int_{C'} \varphi\, ds$$

$$= \int_{C} \varphi\, ds = \Phi(C)$$

by Stokes's Theorem, so C is Φ-minimizing. If Φ is strictly convex, the inequality is strict unless C' is also a straight line from A to B, so C is uniquely minimizing.

10.3. Proposition. *A nonnegative homogeneous C^2 function Φ on $\mathbf{R}^n$ is convex (respectively, uniformly convex) if and only if the restrictions $\Phi(\theta)$ of Φ to circles about the origin satisfy*

$$\Phi''(\theta) + \Phi(\theta) \leq 0 \quad (< 0).$$

Proof. Since convexity in every plane through 0 is equivalent to convexity, we may assume $n = 2$. The curvature κ of any graph $r = f(\theta)$ in polar coordinates is given by

$$\kappa = \frac{f^2 - f f'' + 2f'^2}{(f^2 + f'^2)^{3/2}}.$$

Therefore the curvature of the boundary of the unit ball $r = 1/\Phi(\theta)$ is given by

$$\kappa = \left(\frac{\Phi}{\sqrt{\Phi^2 + \Phi'^2}}\right)^3 (\Phi + \Phi'').$$

The proposition follows.

10.4. Generalized curvature. *Let C be a C^2 curve with arc length parametrization $f : [0, 1] \to \mathbf{R}^n$ and curvature vector κ. Let Φ be a C^2 norm. Consider variations δf supported in $(0, 1)$. Then the first variation satisfies*

$$\delta\Phi(f) = -\int_{[0,1]} D^2\Phi(\kappa) \cdot \delta f\, ds,$$

where $D^2\Phi$ represents the second derivative matrix evaluated at the unit tangent vector **T**. *In particular, for the case of length* $(\Phi(x) = L(x) = |x|)$,

$$\delta L(f) = - \int_{[0,1]} \kappa \cdot \delta f \, ds.$$

In general, we call $D^2\Phi(\kappa)$ the *generalized Φ-curvature vector.*

Proof. Since $\Phi(f) = \int \Phi(f'(u))du$ for any parameterization $f(u)$,

$$\delta\Phi(f) = \int D\Phi(f') \cdot \delta f'(u) du$$

$$= - \int D^2\Phi(f')(f'') \cdot \delta f(u) du$$

by integration by parts. Since for the initial arc length parametrization, f' is the unit tangent vector and f'' is the curvature vector κ, initially

$$\delta\Phi(f) = - \int D^2\Phi(\kappa) \cdot \delta f(s) ds.$$

Remarks. $D^2\Phi(\mathbf{T}) = 0$, and hence the generalized Φ-curvature vector $D^2\Phi(\kappa)$ is normal to **T**. In $\mathbf{R}^2$, $D^2\Phi(\kappa)$ is a multiple of κ: if $\Phi(v) = |v|h(\theta)$, then $D^2\Phi(\kappa) = (h + h'')\kappa$.

10.5. The isoperimetric problem. One famous isoperimetric theorem says that among all closed curves C in $\mathbf{R}^n$ of fixed length, the circle bounds the most area—that is, the oriented area-minimizing surface S of greatest area (see, for example, [Fed, 4.5.14]). In other words, an area-minimizing surface S with given boundary C satisfies

$$\text{area } S \leq \frac{1}{4\pi}(\text{length} C)^2.$$

Given a convex norm Φ or $\mathbf{R}^n$, we seek a closed curve C_0 of fixed cost $\Phi(C_0)$ which bounds the most area, so any area-minimizing surface

S with given boundary C satisfies

$$\text{area} S \le \alpha[\Phi(C)]^2,$$

with equality for $C = C_0$.

Almgren's methods (see Almgren [Alm, especially Section 9]) using geometric measure theory show that such an *optimal isoperimetric curve* exists.

In the plane such curves have a nice characterization. Let Ψ be a 90° rotation of Φ, so

$$\Phi(C) = \int_C \Psi(\mathbf{n}),$$

where $\mathbf{n}$ is the unit normal obtained by rotating the unit tangent $\mathbf{T}$ 90° counterclockwise. The dual norm Ψ^* is defined by

$$\Psi^*(w) = \sup\{v \cdot w : \Psi(v) \le 1\},$$

so $|v \cdot w| \le \Psi(v)\Psi(w)$. The optimal isoperimetric curve is simply the boundary of the unit Ψ^*-ball or "Wulff shape" (see Wulff 1901 [Wul] or Taylor [Tay1]). M. Gromov traces the general result to H. Brunn's inaugural dissertation (see Brunn 1887 [Bru]). Here we sketch a short modern proof, based on Schwartz symmetrization, as used recently by Gromov [Gro] (or see Berger [Ber, 12.11.4]) and earlier by Knothe [Kno]. The same result and proof hold for optimal isoperimetric hypersurfaces in all dimensions.

10.6. Theorem. *Let Ψ be a norm on $\mathbf{R}^2$. Among all curves enclosing the same area, the boundary of the unit Ψ^*-ball B (Wulff shape) minimizes $\int_{\partial B} \Psi(\mathbf{n})$.*

Proof sketch. Consider any planar curve enclosing a region B' of the same area as B. Let f be an area-preserving map from B' to B carrying vertical lines linearly to vertical lines. Then $\det Df = 1$ and Df is triangular:

$$Df = \begin{bmatrix} a & 0 \\ * & b \end{bmatrix}.$$

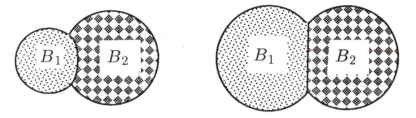

Figure 10.3. Undergraduate students proved recently that these "double bubbles" provide the least-perimeter way to enclose and separate two prescribed areas. Their work led to the recent computer proof by Hass and Schlafly that the similar double soap bubble in $\mathbf{R}^3$ is the least-area way to enclose and separate two equal volumes (Foisy et al. [Foi, Fig. 1.0.1)]).

Since $det Df = ab = 1$, div $f = a + b \geq 2$. Hence

$$
\Psi(\partial B') = \int_{\partial B'} \Psi(\mathbf{n}) \geq \int_{\partial B'} \Psi(\mathbf{n})\Psi^*(f)
$$

$$
\geq \int_{\partial B'} f \cdot \mathbf{n} = \int_{B'} \operatorname{div} f \geq 2 \text{ area } B' = 2 \text{ area } B,
$$

with equality if $B' = B$.

Remarks. Careful attention to the inequalities in the proof recovers the result of J. Taylor [Tay2] that the Wulff shape is the unique minimizer among measurable sets (see Brothers and Morgan [Bro]). The generalized curvature vector of the Wulff shape is just the inward unit normal. The same argument proves that the cap of a Wulff shape inside a convex cone is Φ-minimizing among surfaces enclosing fixed volume inside the cone.

10.7. Double salt crystals. The double cluster problem seeks the least expensive way to enclose and separate two regions of prescribed areas. For the case of length ($\Phi(x) = |x|$), the solution is the "double bubble" of Figure 10.3, as proved quite recently by the 1990 Williams College "SMALL" Undergraduate Research Geometry Group (Foisy et al. [Foi], see also Morgan [Mor5]). Their work led to the recent computer proof

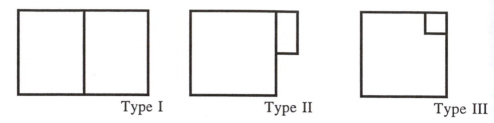

Figure 10.4. Other students found these three different optimal double crystal types for their mathematical model (French et al. [Fre], [Mor8, Figure 1]).

by Hass and Schlafly [Has1], [Has3], [Has4] that the similar double soap bubble in $\mathbf{R}^3$ is the least-area way to enclose and separate two equal volumes (see also Morgan [Mor3] and Peterson [Pet]). For the case $\Phi(x, y) = |x| + |y|$, for which the Wulff shape (ideal single enclosure) is a square, the solutions fall into the *three different types* of Figure 10.4, as proved by the 1993 SMALL Geometry Group (see French et al. [Fre], [Mor8]). Electron microscope photographs of table salt (Figure 10.5) show the same qualitative behavior. If the separating interface counts just some fraction $0 < \lambda < 1$, a similar result holds, except that if $\lambda < \lambda_0 \approx .564$, type II does not occur, as proved by the 1995 SMALL Geometry Group (see Wecht, Barber, and Tice [Wec]).

In general dimensions, optimal isoperimetric curves are not well understood. Obvious candidates are planar Wulff shapes. The following new theorem says, however, that optimal isoperimetric curves are not generally planar.

10.8. Theorem. *For some convex norms Φ in $\mathbf{R}^3$, an optimal isoperimetric curve is nonplanar.*

Proof. Define Φ by taking the unit Φ-ball to be the centrally symmetric polyhedron B of Figure 10.6. In any plane P, which we may translate to pass through the origin, the Wulff shape S with boundary C maximizes (area S) / $\Phi(C)^2$. We will show that this ratio is larger for some nonplanar curve C.

Figure 10.5. Electron microscope photographs of table salt agree qualitatively with the student predictions (French et al. [Fre], [Mor8]).

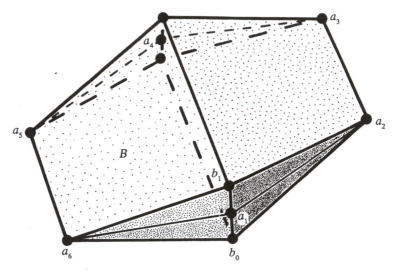

Figure 10.6. The unit Φ-ball B. Any slice A by a plane P
through the origin has a vertex a_1 that is not a vertex of B.

The slice A of the unit Φ-ball B by the plane P must be polygonal. At
least one vertex a_1 is not a vertex of B. The vertex a_1 must lie on an edge
of B, with vertices b_0, b_1. The Wulff shape S is the polygon formed dual
to A, rotated clockwise 90°. (See Figure 10.7.) If C_1 denotes the edge
dual to a_1 (rotated 90°), then C_1 points in the a_1 direction. Its distance
from the origin is $1/|a_1|$.

Let C' be the polygon in $\mathbf{R}^3$ obtained from C by replacing C_1 by two
segments in the directions b_0, b_1, in the order that keeps the projection
PC' of C' onto P out of the interior of C. (See Figure 10.8.) Then
$\Phi(C') = \Phi(C)$. Let S' be an area-minimizing surface bounded by C'.
Then

$$\text{area } S' > \text{area } PS' > \text{area } S.$$

Consequently,

$$\frac{\text{area } S'}{\Phi(C')^2} > \frac{\text{area } S}{\Phi(C)^2}.$$

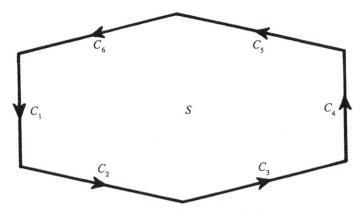

Figure 10.7. In the plane P, the Wulff crystal S is the polygon dual to A, rotated $90°$ clockwise.

Remark. By approximation one obtains examples that are also smooth and elliptic.

For length, optimal isoperimetric curves are circles of constant curvature. For general Φ, the generalized curvature at least satisfies an inequality.

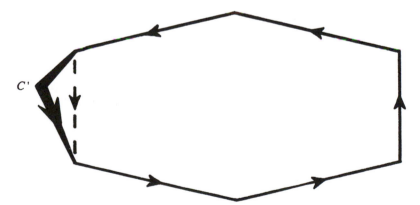

Figure 10.8. Obtain C' from C by replacing C_1 with segments in the directions b_0, b_1. Then $\Phi(C') = \Phi(C)$, but area $S' >$ area S.

10.9. Lemma. *For a C^2 optimal isoperimetric curve C_0, the generalized Φ-curvature vector satisfies*

$$|D^2\Phi(\kappa)| \leq \frac{\Phi(C_0)}{2 \text{ area } S_0}. \tag{10.1}$$

Remarks. For the case where Φ is length and C_0 is the unit circle, (10.1) says $|\kappa| \leq 1$. The smoothness hypothesis on C_0 is unnecessary; still the conclusion implies that C_0 is $C^{1,1}$. If C_0 bounds a unique smooth area-minimizing surface S_0 with $\mathbf{n}$ the inward normal to C_0 along S_0, $D^2\Phi(\kappa)$ actually must be a constant multiple of $\mathbf{n}$. In particular, a planar optimal isoperimetric curve has constant generalized curvature:

$$|D^2\Phi(\kappa)| = K.$$

Proof. Let $f : [0, a] \to \mathbf{R}^n$ be a local arc length parameterization of C_0. Consider compactly supported variations δf. Then

$$
\begin{aligned}
0 \geq\ & \delta(\text{area } S - \alpha\Phi(C)^2) \\
\geq\ & -\int |\delta f| ds + 2\alpha\Phi(C_0) \int D^2\Phi(\kappa) \cdot \delta f\, ds
\end{aligned}
$$

by the Generalized Curvature Proposition 10.4. Therefore

$$|D^2\Phi(\kappa)| \leq \frac{1}{2\alpha\Phi(C_0)} = \frac{\Phi(C_0)}{2 \text{ area } S_0}.$$

A norm Φ is called *crystalline* if the unit Φ-ball is a polytope.

10.10. Conjecture. *If Φ is crystalline, then an optimal isoperimetric curve is a polygon.*

10.11. Φ-minimizing networks. A *network N* is a finite collection of line segments. Given a norm Φ and a finite set of boundary points in $\mathbf{R}^n$, we seek a Φ-minimizing network connecting the points. For the case where Φ is length, the generalized "Steiner" or "Fermat" problem (see Fermat, 1638, [Fer, p. 153]; Steiner 1835, 1837 [Ste1], [Ste2]; Jarnik and Kossler 1934 [Jar]), such networks meet only in threes at equal $120°$

Figure 10.9. Length-minimizing networks meet in threes at equal angles of 120°.

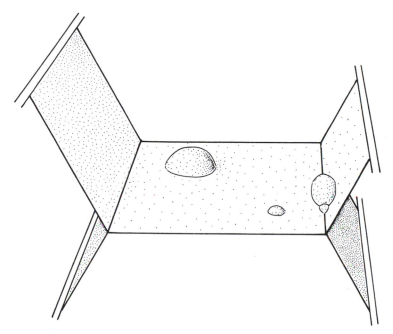

Figure 10.10. Soap films meet in threes at 120° angles in an attempt to minimize area.

angles (or in twos at boundary points at angles of at least 120°), as shown in Figure 10.9. Soap film strips behave similarly in their quest to minimize area, as shown in Figure 10.10 (see also Courant and Robbins [Cou, pp. 354–361, 392]).

Recently there have appeared results on general norms Φ, many by undergraduates. (See the surveys by Morgan [Mor6], [Mor7].)

10.12. Theorem. (Levy [Lev2], Alfaro et al. [Alf2]). *Let Φ be a differentiable, uniformly convex norm on $\mathbf{R}^2$. Then Φ-minimizing networks meet only in threes.*

The proof shows that a junction of four or more segments is unstable.

10.13. Theorem. (Lawlor and Morgan [Law, Theorem 4.4]). *Let Φ be a differentiable norm on $\mathbf{R}^n$. In Φ-minimizing networks, $n+1$ segments can meet at a point, but never $n+2$.*

It turns out that all such junctions locally can be "calibrated" and classified.

The next theorem, in comparison with Theorem 10.12, exhibits a surprising sensitivity to smoothness class.

10.14. Theorem. (1988 SMALL Geometry Group, Alfaro et al. [Alf1], [Alf2]). *Consider piecewise differentiable, uniformly convex norms Φ on $\mathbf{R}^2$. Then Φ-minimizing networks sometimes meet in fours, although never in fives.*

The proof shows that an X can be Φ-minimizing by symmetry arguments and calculus. The proof is much easier for the "rectilinear norm" or "Manhattan metric" Φ_M, which is not *uniformly* convex (see Hanan [Han]).

E. Cockayne [Coc] earlier studied planar norms, but did not discuss dependence on differentiability.

10.15. Theorem. (Conger [Con]). *Consider piecewise differentiable, uniformly convex norms Φ on $\mathbf{R}^3$. Then Φ-minimizing networks sometimes meet in sixes.*

The proof shows that six segments meeting along orthogonal axes in $\mathbf{R}^3$ are Φ-minimizing for some Φ. The large number of possible competitors requires cleverness as well as persistence in the proof.

Conger conjectured that the sharp bound for the number of segments meeting in a Φ-minimizing network in $\mathbf{R}^n$ is $2n$.

For non-uniformly convex norms, if the unit Φ-ball is a cube in $\mathbf{R}^n$, the network consisting of the 2^n segments from the center to the vertices is Φ-minimizing, with an easy proof. This probably exhibits the upper bound (see Füredi, Lagarias, and Morgan [Für, Introduction, Theorem 2.1]).

See related work by D. Cieslik [Cie1], [Cie2] and K. Swanepoel [Swa].

Selected Formulas

Curvature vector:

$$\boldsymbol{\kappa} = d\mathbf{T}/ds \tag{2.1}$$

Arc length of curve $u(t)$:

$$\int \sqrt{\sum g_{ij} \dot{u}^i \dot{u}^j}\, dt \tag{3.1}$$

$(\dot{u}^i = du^i/dt)$

Metric of surface $\mathbf{x}(u^i)$:

$$g_{ij} = \mathbf{x}_i \cdot \mathbf{x}_j \tag{3.2}$$

$(u^1, u^2, \ldots$ give parameters on surface; $\mathbf{x}_i = \partial \mathbf{x}/\partial u^i)$

Inverse matrix g^{ij}

2-dimensional surface $\mathbf{x}(u^1, u^2)$ in $\mathbf{R}^3$:
  Mean curvature $H = \text{trace II} = \kappa_1 + \kappa_2$

$$H = \frac{\mathbf{x}_2^2 \mathbf{x}_{11} - 2(\mathbf{x}_1 \cdot \mathbf{x}_2)\mathbf{x}_{12} + \mathbf{x}_1^2 \mathbf{x}_{22}}{\mathbf{x}_1^2 \mathbf{x}_2^2 - (\mathbf{x}_1 \cdot \mathbf{x}_2)^2} \cdot \mathbf{n} \tag{3.4}$$

Gauss curvature $G = \det \text{II} = \kappa_1 \kappa_2$

$$G = \frac{(\mathbf{x}_{11} \cdot \mathbf{n})(\mathbf{x}_{22} \cdot \mathbf{n}) - (\mathbf{x}_{12} \cdot \mathbf{n})^2}{\mathbf{x}_1^2 \mathbf{x}_2^2 - (\mathbf{x}_1 \cdot \mathbf{x}_2)^2} \tag{3.5}$$

For the graph of a function $f : \mathbf{R}^2 \to \mathbf{R}$,

$$H = \frac{(1 + f_y^2)f_{xx} - 2f_x f_y f_{xy} + (1 + f_x^2)f_{yy}}{(1 + f_x^2 + f_y^2)^{3/2}} \tag{3.6}$$

$$G = \frac{f_{xx}f_{yy} - f_{xy}^2}{(1 + f_x^2 + f_y^2)^2} \tag{3.7}$$

2-dimensional surface $\mathbf{x}(u^1, u^2)$ in $\mathbf{R}^n$:

Mean curvature vector $\mathbf{H} = \text{trace II}$

$$\mathbf{H} = P\frac{\mathbf{x}_2^2 \mathbf{x}_{11} - 2(\mathbf{x}_1 \cdot \mathbf{x}_2)\mathbf{x}_{12} + \mathbf{x}_1^2 \mathbf{x}_{22}}{\mathbf{x}_1^2 \mathbf{x}_2^2 - (\mathbf{x}_1 \cdot \mathbf{x}_2)^2} \tag{4.1}$$

Gauss curvature $G = \det \text{II}$

$$G = \frac{(P\mathbf{x}_{11}) \cdot (P\mathbf{x}_{22}) - (P\mathbf{x}_{12})^2}{\mathbf{x}_1^2 \mathbf{x}_2^2 - (\mathbf{x}_1 \cdot \mathbf{x}_2)^2} \tag{4.2}$$

where P denotes projection onto $T_p S^\perp$.
For the graph of a function $f : \mathbf{R}^{n-1} \to \mathbf{R}$,

$$H = \text{div}\frac{\Delta f}{\sqrt{1 + |\nabla f|^2}} = \frac{(1 + |\nabla f|^2)\Delta f - \sum f_i f_j f_{ij}}{(1 + |\nabla f|^2)^{3/2}},$$

where $f_i = \partial f/\partial x_i, f_{ij} = \partial^2 f/\partial x_i \partial x_j, \nabla f = (f_1, \ldots, f_{n-1})$, div $(p, q, \ldots) = p_1 + q_2 + \cdots$, and $\nabla f = \text{div } \nabla f = f_{11} + f_{22} + \cdots$ (Exercise 5.12.).

For the level set $\{f = c\}$ of a function $f : \mathbf{R}^n \to \mathbf{R}$,

$$H = -\text{div}\frac{\nabla f}{|\nabla f|} \tag{5.1}$$

Christoffel symbols:

$$\Gamma^i_{jk} = \frac{1}{2}\sum_l g^{il}(g_{lj,k} + g_{lk,j} - g_{jk,l}) \qquad (6.2)$$

Riemannian curvature tensor:

$$R^i_{jkl} = a_{ik} \cdot a_{jl} - a_{jk} \cdot a_{il} \qquad (5.4)$$

$$R^i_{jkl} = -\Gamma^i_{jk,l} + \Gamma^i_{jl,k} + \sum_h(-\Gamma^h_{jk}\Gamma^i_{hl} + \Gamma^h_{jl}\Gamma^i_{hk}). \qquad (6.4)$$

(a_{jk} are components of second fundamental form II in orthonormal coordinates.)

Ricci curvature:

$$R_{jl} = \sum_i R^i_{jil} \qquad (6.7)$$

Scalar curvature:

$$R = \sum g^{jl} R_{jl} \qquad (6.8)$$

(If S is 2-dimensional, $G = R/2$.)

Sectional curvature for v, w orthonormal:

$$\begin{aligned} K(v \wedge w) &= \text{II}(v,v) \cdot \text{II}(w,w) - \text{II}(v,w)^2 \qquad (5.2)\\ &= \sum g_{ih} R^h_{jkl} v^i w^j v^k w^l. \qquad (6.9) \end{aligned}$$

Covariant derivative of a vectorfield X^i:

$$X^i_{;j} = X^i_{,j} + \sum_k \Gamma^i_{jk} X^k \qquad (6.1)$$

Geodesics $u(t), t$ arc length:

$$0 = \ddot{u}^i + \sum_{j,k} \Gamma_{jk}^i \dot{u}^j \dot{u}^k \tag{6.17}$$

Gradient:

$$\nabla f^i = g^{ij} f_{,j}$$

Laplacian:

$$\triangle f = g^{ij}(f_{,ij} - \Gamma_{ij}^k f_{,k}) = \frac{1}{\sqrt{\det g}} \frac{\partial}{\partial u_i}(\sqrt{\det g}\; g^{ij} f_{,j})$$

Solutions to
Selected Exercises

2.1 a. $\kappa(0) = (0, 14)$

b. $\kappa(0) = 14/(1 + 36)^{3/2} = 14/37^{3/2}$. Since the curve has slope 6, the unit normal $\mathbf{n}$ has slope $-1/6$, $\mathbf{n} = (-6, 1)/\sqrt{37}$, $\kappa(0) = 14\,(-6, 1)/\sqrt{37}$

c. same as **b.**

2.2 b. $\kappa = (1+t)^{-3} \left(-2s - c - tc - ts - t^2c,\ 2c - s - ts + tc - t^2s,\ t^{-1/2}/\sqrt{2} - t^{1/2}/\sqrt{2}\right)$, where $s = \sin t$, $c = \cos t$. Note that for t large, $\kappa \sim -(c, s, 0)/t$ is small and points towards the z axis.

c. 12 feet.

3.1 Curvature in the direction θ is $2\cos\theta\sin\theta$. The principal directions are $\pi/4$, $3\pi/4$, $\kappa_1 = 1$, $\kappa_2 = -1$, $H = 0$, $G = -1$, $\kappa_1' + \kappa_2' = 0$, $\kappa_1'\kappa_2' = 0$.

3.2 a. $\kappa_1 = 1/a$, $\kappa_2 = 1/a$, $G = 1/a^2$, $H = 2/a$ (or $-1/a, -1/a, 1/a^2, -2/a$)

b. $2a$, $2b$, $4ab$, $2(a + b)$

c. $\text{II} = D^2 f(0, 0) = \begin{bmatrix} 132 & -24 \\ -24 & 118 \end{bmatrix}$, $\quad H = 132 + 118 = 250$,

$G = 132 \cdot 118 - 24^2 = 15{,}000$, $\quad \kappa = \frac{H \pm \sqrt{H^2 - 4G}}{2} = 100,\ 150$.

d. Note that the x, y-plane is not the tangent plane at $\mathbf{0}$, so use Equations 3.6 and 3.7. $H = 4\sqrt{2}$, $G = 6$. Hence, $\kappa = 3\sqrt{2}, \sqrt{2}$.

e. Switch variables and use Equations 3.6 and 3.7. $H = 0$, $\quad G = -1/(1 + y^2 \sec^2 z)^2 = -1/(1 + x^2 + y^2)^2$, $\quad \kappa = \pm 1/(1 + y^2 \sec^2 z) = \pm 1/(1 + x^2 + y^2)$.

f. Use Equations 3.6 and 3.7 and implicit differentiation.

$$H = \frac{5 \cdot 3^4 x^2 + 10 \cdot 2^4 y^2 + 13 z^2}{(81 x^2 + 16 y^2 + z^2)^{3/2}},$$

$$G = \left(\frac{36}{81 x^2 + 16 y^2 + z^2} \right)^2,$$

$$\kappa = \frac{H \pm \sqrt{H^2 - 4G}}{2}.$$

3.4 $2\pi a \sin c$. (It is a circle of radius $a \sin c$.)

3.6 $\mathbf{x} = (x, y, f(x, y))$, $\mathbf{n} = \dfrac{(-f_x, -f_y, 1)}{\sqrt{1 + f_x^2 + f_y^2}}$,

$$H = \frac{(1 + f_y^2) f_{xx} - 2 f_x f_y f_{xy} + (1 + f_x^2) f_{yy}}{(1 + f_x^2)(1 + f_y^2) - f_x^2 f_y^2} \frac{1}{\sqrt{1 + f_x^2 + f_y^2}},$$

$$G = \frac{f_{xx} f_{yy} - f_{xy}^2}{1 + f_x^2 + f_y^2} \frac{1}{1 + f_x^2 + f_y^2}.$$

3.7 $\mathbf{x}_z = (f' \cos\theta, f' \sin\theta, 1)$, $\mathbf{x}_\theta = (-f \sin\theta, f \cos\theta, 0)$,

$\mathbf{n} = -(\cos\theta, \sin\theta, -f')(1 + f'^2)^{-1/2}$ (inward),

$\mathbf{x}_{zz} = (f'' \cos\theta, f'' \sin\theta, 0)$, $\mathbf{x}_{z\theta} = (-f' \sin\theta, f' \cos\theta, 0)$,

$\mathbf{x}_{\theta\theta} = (-f \cos\theta, -f \sin\theta, 0)$,

$$H = \frac{-f^2 f'' + f(1 + f'^2)}{f^2(1 + f'^2)} \frac{1}{\sqrt{1 + f'^2}} = \kappa + \frac{1}{f\sqrt{1 + f'^2}}.$$

3.9 0, by Theorem 3.2, because $H = 0$.

3.10 $A = \displaystyle\int_{\varphi=0}^{r/a} 2\pi a \sin\varphi \, a \, d\varphi = 2\pi a^2 \left(1 - \cos\frac{r}{a} \right) = 2\pi r^2 - \frac{1}{a^2} \frac{\pi}{12} r^4 + \dots$

4.1 $(-4/9, 4/9, 2/9, 0)$

4.2 a. $\mathbf{x} = (x, y, x^2 + 2y^2, 66x^2 - 24xy + 59y^2)$, $\mathbf{x}_1 = (1, 0, 0, 0)$, $\mathbf{x}_2 = (0, 1, 0, 0)$, $\mathbf{x}_{11} = (0, 0, 2, 132)$, $\mathbf{x}_{12} = (0, 0, 0, -24)$, $\mathbf{x}_{22} = (0, 0, 4, 118)$. $P\mathbf{x}_{ij} = \mathbf{x}_{ij}$. By Proposition 4.2, $\mathbf{H} = (0, 0, 6, 250)$, $G = 15,008$. Note that answers are sums of answers to Ex. 3.2 (b, c).

b. $\mathbf{x} = (x, y, x^2 - y^2, 2xy)$, $\mathbf{x}_1 = (1, 0, 2x, 2y)$, $\mathbf{x}_2 = (0, 1, -2y, 2x)$, $\mathbf{x}_{11} = (0, 0, 2, 0)$, $\mathbf{x}_{12} = (0, 0, 0, 2)$, $\mathbf{x}_{22} = (0, 0, -2, 0)$.

By Proposition 4.2, $\mathbf{H} = P\dfrac{0}{\text{something}} = 0$. For G, we need $P\mathbf{x}_{ij}$.

Since $\mathbf{x}_1, \mathbf{x}_2$ are orthogonal,

$$P(\mathbf{x}_{11}) = \mathbf{x}_{11} - \frac{\mathbf{x}_{11} \cdot \mathbf{x}_1}{\mathbf{x}_1^2} \mathbf{x}_1 - \frac{\mathbf{x}_{11} \cdot \mathbf{x}_2}{\mathbf{x}_2^2} \mathbf{x}_2 = \dots = \frac{(-4x, 4y, 2, 0)}{1 + 4x^2 + 4y^2},$$

$$P(\mathbf{x}_{12}) = \frac{(-4y, -4x, 0, 2)}{1 + 4x^2 + 4y^2}, \quad P(\mathbf{x}_{22}) = \frac{(4x, -4y, -2, 0)}{1 + 4x^2 + 4y^2}.$$

$$G = -8(1 + 4x^2 + 4y^2)^{-3} = -8(1 + 4|z|^2)^{-3}.$$

4.3 a. $y_1 = \sqrt{1 - x_1^2}$, $y_2 = \sqrt{1 - x_2^2}$

$$\mathrm{II}(0, 1, 0, 1) = \begin{bmatrix} \begin{bmatrix} -1 \\ 0 \end{bmatrix} & \mathbf{0} \\ \mathbf{0} & \begin{bmatrix} 0 \\ -1 \end{bmatrix} \end{bmatrix}$$

Therefore $H = \sqrt{2}, G = 0$.

4.4 $\mathbf{x} = (z, f(z))$, with $z = u_1 + iu_2$.

$\mathbf{x}_1 = (1, f'(z))$, $\mathbf{x}_2 = (i, if'(z))$, $\mathbf{x}_1 \cdot \mathbf{x}_2 = 0$ ($\mathbf{x}_2 = i\mathbf{x}_1$) ,

$\mathbf{x}_{11} = (0, f''(z))$, $\mathbf{x}_{12} = (0, if''(z))$, $\mathbf{x}_{22} = (0, -f''(z))$.

$$\mathbf{H} = \frac{P(0)}{\text{something}} = 0. \text{ For } G, \text{ we need } P\mathbf{x}_{ij}.$$

Since $\mathbf{x}_1, \mathbf{x}_2$ constitute a basis over $\mathbf{R}$ for the complex subspace spanned by $(1, f'(z))$ over $\mathbf{C}$, P just projects onto the orthogonal complex subspace spanned by $(-\overline{f'(z)}, 1) \in \mathbf{C}^2$.

$$P(\mathbf{x}_{11}) = -P(\mathbf{x}_{22}) = \frac{(-\overline{f'}f'', f'')}{(1 + |f'|^2)},$$

$$P(\mathbf{x}_{12}) = \frac{(-i\overline{f'}f'', if'')}{(1 + |f'|^2)},$$

$$G = -2|f''|^2(1 + |f'|^2)^{-3}.$$

4.5 Calculating with formula 4.1 yields

$\mathbf{H} = P\mathbf{v}(1 + |f_x|^2 + |f_y|^2)^{-1}$, where

$\mathbf{v} = (1 + |f_y|^2)f_{xx} - 2(f_x \cdot f_y)f_{xy} + (1 + |f_x|^2)f_{yy}$

$\in \mathbf{R}^{n-2} \subset \mathbf{R}^2 \times \mathbf{R}^{n-2}$.

Clearly if $\mathbf{v} = 0$, then $\mathbf{H} = 0$ and the surface is minimal.

On the other hand, if $\mathbf{H} = 0, \mathbf{v} \in \ker P \cap \mathbf{R}^{n-2} = \{0\}$.

5.1 $H = 18$

5.2 $-6e_{134} - 12e_{234}$

5.3 One possibility is

$$
\begin{aligned}
u &= (-1, 0, 1, -1)\\
v &= (0, -1, 1, -1),\\
w &= \left(-\frac{1}{\sqrt{3}}, 0, \frac{1}{\sqrt{3}}, -\frac{1}{\sqrt{3}}\right).\\
z &= \left(\frac{2}{\sqrt{15}}, -\frac{3}{\sqrt{15}}, \frac{1}{\sqrt{15}}, -\frac{1}{\sqrt{15}}\right).\\
u \wedge v &= e_{12} - e_{13} + e_{14} + e_{23} - e_{24}\\
w \wedge z &= \frac{1}{\sqrt{5}} e_{12} - \frac{1}{\sqrt{5}} e_{13} + \frac{1}{\sqrt{5}} e_{14} + \frac{1}{\sqrt{5}} e_{23} - \frac{1}{\sqrt{5}} e_{24}\\
&= \frac{1}{\sqrt{5}} u \wedge v\\
|w \wedge z| &= 1.
\end{aligned}
$$

5.5 $e_{12} + 2e_{13} + 2e_{23} = (e_1 + e_2) \wedge (e_2 + 2e_3)$.

5.6 Method 1: Assume $e_{12} + e_{34} = (\sum a_i e_i) \wedge (\sum b_j e_j)$, and derive a contradiction.
Method 2: Clearly, if ξ is simple, $\xi \wedge \xi = 0$. Since $(e_{12} + e_{34}) \wedge (e_{12} + e_{34}) = 2e_{1234} \neq 0$, it is not simple. (Actually, for $\xi \in \Lambda_2 \mathbf{R}^n$, ξ simple $\Leftrightarrow \xi \wedge \xi = 0$. For $\xi \in \Lambda_m \mathbf{R}^n$, $m > 2$, ξ simple $\Rightarrow \xi \wedge \xi = 0$.)

5.7 For $\mathbf{S}^m$, the slices are all spheres with Gauss curvature $1/a^2$, so the scalar curvature is just $m(m-1)/a^2$. For $\mathbf{S}^2(a) \times \mathbf{S}^3(b) \subset \mathbf{R}^3 \times \mathbf{R}^4 = \mathbf{R}^7$, each slice is either a sphere or a torus and the sectional curvatures are all $1/a^2, 1/b^2$, or 0, yielding a scalar curvature of $2/a^2 + 6/b^2$.

5.8
$$
\begin{array}{lll}
R_{1212} = 3, & R_{1213} = -1, & R_{1223} = -5,\\
R_{1312} = -1, & R_{1313} = -4, & R_{1323} = 0,\\
R_{2312} = -5, & R_{2313} = 0, & R_{2323} = -1;
\end{array}
$$
The Ricci curvature is the symmetric matrix
$$
\begin{bmatrix}
-1 & 0 & 5\\
0 & 2 & -1\\
5 & -1 & -5
\end{bmatrix}
$$
For example, $R_{11} = R_{2121} + R_{3131} = 3 - 4 = -1$.

5.9 a. $\mathrm{II} = \left[\frac{\partial^2 y}{\partial x^2}\right]_0 = \begin{bmatrix} (2,6) & (2,0) & (0,0)\\ (2,0) & (2,2) & (0,2)\\ (0,0) & (0,2) & (10,2) \end{bmatrix}$.

b. $K(e_1 \wedge e_2) = (2,6) \cdot (2,2) - (2,0) \cdot (2,0) = 12$.

c. One orthonormal basis is $v = (1, -1, 0)/\sqrt{2}$, $w = (0, 0, 1)$.
$$
\begin{aligned}
K &= \mathrm{II}(v,v) \cdot \mathrm{II}(w,w) - \mathrm{II}(v,w) \cdot \mathrm{II}(v,w)\\
&= (0,4) \cdot (10,2) - (0, -\sqrt{2}) \cdot (0, -\sqrt{2}) = 6.
\end{aligned}
$$

One orthonormal basis for $\{x_1 + x_2 + x_3 = 0\}$ is

$$v = \frac{(1,-1,0)}{\sqrt{2}}, \quad w = \frac{(1,1,-2)}{\sqrt{6}}. \quad K = 0.$$

d. $R_{1212} = 12, R_{1213} = 12, R_{1223} = 0, R_{1313} = 32, R_{1323} = 20, R_{2323} = 20.$
Rest by symmetries. Redoing b and c, we have the following:

b. $R_{1212} = 12.$

c. $v_1 = -v_2 = 1/\sqrt{2}, w_3 = 1,$ rest 0.

$K = \frac{1}{2}R_{1313} - \frac{1}{2}R_{1323} - \frac{1}{2}R_{2313} + \frac{1}{2}R_{2323} = 6.$

$v_1 = -v_2 = 1/\sqrt{2}, v_3 = 0, w_1 = w_2 = 1/\sqrt{6}, w_3 = -2/\sqrt{6}.$

$K = \frac{1}{12}R_{1212} - \frac{1}{6}R_{1213} - \frac{1}{6}R_{1312} + \frac{1}{3}R_{1313} - \frac{1}{12}R_{1221} + \frac{1}{6}R_{1321}$

$\quad - \frac{1}{3}R_{1323} - \frac{1}{12}R_{2112} + \frac{1}{6}R_{2113} - \frac{1}{3}R_{2313} + \frac{1}{12}R_{2121} + \frac{1}{3}R_{2323}$

$= 1 - 2 - 2 + 32/3 + 1 - 2 - 20/3 + 1 - 2 - 20/3 + 1 + 20/3 = 0.$

e. Ric $= \begin{bmatrix} 44 & 20 & 0 \\ 20 & 32 & 12 \\ 0 & 12 & 52 \end{bmatrix}, R = 44 + 32 + 52 = 128.$

5.10 a. $w = f(x, y, z) = -\sqrt{a^2 - x^2 - y^2 - z^2}$

b. $f_{xx}(0) = 1/a, f_{xy}(0) = 0,$ etc.

c. $H = 3/a, \mathbf{H} = (3/a)e_4$ ⠀⠀(normal to S at p)

d. $\mathrm{II}(v, v) = v \cdot ((1/a)\mathrm{I})(v) = (1/a)v \cdot v = 1/a.$

$\mathrm{II}(v, w) = v \cdot (1/a)\mathrm{I}(w) = (1/a)v \cdot w = 0.$

Hence $K(v \wedge w) = (1/a)(1/a) - 0^2 = 1/a^2.$

e. $R_{ijij} = 1/a^2 \ (i \neq j),$ rest 0. For example, $R_{1223} = \begin{vmatrix} 0 & 0 \\ 1/a & 0 \end{vmatrix} = 0.$

f. $K(v \wedge w) = \sum R_{ijkl}v^i w^j v^k w^l = \sum_{i \neq j} R_{ijij}v^i w^j v^i w^j - \sum_{i \neq j} R_{ijji}v^i w^j v^j w^i$

$= \frac{1}{a^2}(\sum (v^i)^2)(\sum (w^j)^2) - \frac{1}{a^2}(\sum v^i w^i)(\sum v^j w^j)$

$= \frac{1}{a^2}(1)(1) - \frac{1}{a^2}(0)(0) = \frac{1}{a^2}.$

g. $R_{jl} = \sum R_{ijil} = 0$ if $j \neq l, R_{jj} = \sum R_{ijij} = 2/a^2,$ so $[R_{jl}]$ is just $(2/a^2)\mathrm{I}.$

h. $R = \operatorname{tr} R_{jl} = 6/a^2.$

i. $H = -3/a.$

j. For example, $x_2 = a(-\sin u_1 \sin u_2, \sin u_1 \cos u_2, 0, 0); g_{22} = x_2 \cdot x_2 = a^2 \sin^2 u_1.$

k. $\kappa = (-\cos t, -\sin t, 0, 0)/a$ points inward, so its projection is $\mathbf{0}.$ The length of the path is $2\pi a.$

5.11 The full derivative in $\mathbf{R}^3$

$$Df = \begin{bmatrix} \frac{\partial f}{\partial x} & \frac{\partial f}{\partial y} & \frac{\partial f}{\partial z} \end{bmatrix} = \begin{bmatrix} 0 & 2y & 0 \\ 1 & 0 & 1 \\ 0 & z & y \end{bmatrix} = (0, 2y, 0)\mathbf{i} + (1, 0, 1)\mathbf{j} + (0, z, y)\mathbf{k}.$$

$$Df(0,0,1) = \begin{bmatrix} 0 & 0 & \vdots & 0 \\ 1 & 0 & \vdots & 1 \\ \cdots\cdots\cdots \\ 0 & 1 & & 0 \end{bmatrix} = (0,0,0)\mathbf{i} + (1,0,1)\mathbf{j} + (0,1,0)\mathbf{k}.$$

Restricting to derivatives in the tangent directions (the first two columns) and projecting onto the tangent space (the first two rows) yields the covariant derivative
$$\begin{bmatrix} 0 & 0 \\ 1 & 0 \end{bmatrix} = (1,0)\mathbf{j}.$$

5.12 Let $g(x_1, \ldots, x_n) = x_n - f(x_1, \ldots, x_{n-1})$. Then the graph of f is the level set $\{g = 0\}$. By Equation 5.1,

$$H = -\mathrm{div}\frac{\nabla g}{|\nabla g|} = -\mathrm{div}\frac{(-f_1, \ldots, -f_{n-1}, 1)}{\sqrt{1 + f_1^2 + \cdots + f_{n-1}^2}}$$

$$= \mathrm{div}\frac{\nabla f}{\sqrt{1 + |\nabla f|^2}} - \frac{\partial}{\partial x_n}\frac{1}{\sqrt{1 + |\nabla f|^2}} = \mathrm{div}\frac{\nabla f}{\sqrt{1 + |\nabla f|^2}}.$$

6.2 Write T as the graph of a function over the tangent plane at $p = (0, 1, 0, 1, \ldots)$, namely the $x_1, x_3, \ldots, x_{2n-1}$-plane:
$$(x_2, x_4, \ldots) = f(x_1, x_3, \ldots) = (\sqrt{1 - x_1^2}, \sqrt{1 - x_3^2}, \ldots),$$
and compute that at p

$$\mathrm{II} = D^2 f(0, 0, \ldots) = \begin{bmatrix} \begin{bmatrix} -1 \\ 0 \\ 0 \\ \vdots \end{bmatrix} & 0 & 0 & \cdots \\ 0 & \begin{bmatrix} 0 \\ -1 \\ 0 \\ \vdots \end{bmatrix} & 0 & \cdots \\ \vdots & & & \ddots \end{bmatrix}$$

so that $R_{ijkl} = 0$ and $K = 0$. Now $K = 0$ everywhere by symmetry.

Alternatively, parametrize T as $\mathbf{x} = (\cos\theta_1, \sin\theta_1\cos\theta_2, \sin\theta_2, \ldots)$, compute $[g_{ij}] = [\mathbf{x}_i \cdot \mathbf{x}_j] = \mathrm{I}$ (which is constant!), so the Christoffel symbols, Riemannian curvature, and sectional curvature vanish.

6.3 c. To go from the system to the single differential equation, use $\dot{\varphi} = \frac{d\varphi}{d\theta}\frac{d\theta}{dt} = \varphi'\dot{\theta}$ and hence $\ddot{\varphi} = \varphi''\dot{\theta}^2 + \varphi'\ddot{\theta}$, and then eliminate $\ddot{\theta}$.

6.4 b. $\Gamma^1_{22} = -\cos u_1 \sin u_1 = -\frac{1}{2}\sin 2u_1,\quad \Gamma^1_{33} = \frac{1}{2}\sin 2u_1,$
$\Gamma^2_{12} = \cot u_1,\quad \Gamma^3_{13} = -\tan u_1,$ rest 0.

d. $|\partial/\partial u_1|^2 = g_{11} = a^2,$ so $v = a^{-1}(\partial/\partial u_1).$
$\mathrm{Ric}(v,v) = a^{-2}R_{11} = 2/a^2, m-1 = 2$ times the average $1/a^2$ of curvatures of sections containing v.

e. $R = g^{jl}R_{jl} = 6/a^2, m(m-1) = 6$ times the average of all sectional curvatures.

f. It is not a great circle; it does not lie in a 2-plane.

7.1 3.8"/century. Small ellipticity makes accurate experimental measurement difficult.

7.2 $r = r_0$ and $\varphi = \pi/2$. The first equation for geodesics yields $(d\theta/dt)^2 = GMr_0^{-3}$. The speed $r_0(d\theta/dt) = (GMr_0^{-1})^{1/2}$, exactly analogous to the classical result. The observer must be sufficiently distant to make the relativistic time distortion negligible. Note also that while the circumference of the orbit is $2\pi r_0$ (because there is no relativistic distortion in the tangential direction), the distance to the origin is not r_0 (because there is relativistic distortion in the radial direction, in our coordinates).

7.3 a. yes

b. We're given that $r_0(d\theta/dt) = 3/5; r, \varphi$ constant, with $\varphi = \pi/2$. Therefore $d\tau^2 = -r_0^2 d\theta^2 + dt^2 = (16/9)r_0^2 d\theta^2. \Delta\tau = (4/3)r_0\Delta\theta = (8/3)\pi r_0.$

7.4 Assume φ, θ constant. Now Equations (7.14), (7.18) yield
$-(dr/dt)^2\beta^2(1 - a^2r^{-1})^{-3} + \beta^2(1 - a^2r^{-1})^{-1} = 1.$
The given initial speed implies $\beta^2 = 1$. Substituting w for the small quantity $ar^{-1/2}$ yields $\dfrac{2dw}{w^4(1 - w^2)} = a^{-2}dt.$

By partial fractions $(\frac{2}{w^4} + \frac{2}{w^2} + \frac{1}{1+w} + \frac{1}{1-w})dw = a^{-2}dt.$
Integration yields $2w^{-1} + \frac{2}{3}w^{-3} - \ln|\frac{1+w}{1-w}| = -a^{-2}t + c_0.$
Since $\alpha = \sqrt{2GM} = 1.6293 \times 10^{10} \mathrm{m}^{3/2}/\mathrm{sec}, w_1 = 42,069$ m/sec $= 1.4023 \times 10^{-4}$ (converted using speed of light); $w_2 = 2.0586 \times 10^{-3}$. Now $\Delta t \approx 6.416 \times 10^{31} \mathrm{m}^3/\mathrm{sec}^2 = 2.376 \times 10^6$ sec $\approx 27\frac{1}{2}$ days. As $w \to 1, t \to +\infty.$

8.3 Hint: Given two triangulations, consider a common sub-triangulation.

8.6 The direction vector v moves by parallel transport eastward along a fixed circle of latitude as the earth turns. Since the circle of latitude turns left, v seems to turn right (clockwise), $240° = 4\pi/3$ in a whole day. By Gauss-Bonnet, $2\pi - 4\pi/3$ must equal the integral of the Gauss curvature over the enclosed polar cap, the area of the portion of the unit sphere above latitude L, which equals $2\pi - 2\pi \sin L$ by exercise 3.5. Therefore $\sin L = 2/3$ and L is about $42°$.

8.7 a. True. It suffices to show that any unit tangent vector v at the north pole can be parallel transported to any other unit tangent vector w at the north pole. Just transport it in the direction it points down to the equator, transport it along the equator, and then transport it back up along the meridian where w points.

b. False. A tangent vector at a point on the cylinder cannot be parallel transported to any other tangent vector at that point.

8.8 Although one could write the sphere as a graph, a level set, or a parametrized surface, it is easiest to work from first principles, because the relevant 1-dimensional slices are all circles of radius a.

a. $1/a^2, 1/a^2, 2/a^2, 2/a, (2/a)(-\mathbf{x}/a)$

b. 0

c. $\kappa = 1/(a \sin \varphi), \kappa_g = \cos \varphi/(a \sin \varphi)$

8.9 a. $H = 1/z\sqrt{2}, G = 0$. (Use Equations 3.6 and 3.7, not $\mathrm{II} = D^2 f$, since the x, y-plane is not tangent to C. Note that H and G are not defined at the vertex.)

b. All 0.

c. $\kappa = 1/z$, so $\kappa_g = 1/z\sqrt{2}$.

d. $\pi r^2/\sqrt{2}$.

e. Smoothness fails.

f. $2\sqrt{2} \sin(\pi/2\sqrt{2})$ (As suggested by $G = 0$, the cone can be cut open and bent (without stretching) into a flat planar piece of surface, and the problem is reduced to simple trigonometry.)

Bibliography

Section number follow the entries, indicating the locations in this book where the publications are cited.

[Alf1]   Manuel Alfaro, Mark Conger, Kenneth Hodges, Adam Levy, Rajiv Kochar, Lisa Kuklinski, Zia Mahmood, and Karen von Haam. Segments can meet in fours in energy-minimizing networks. *J. Und. Math.* 22 (1990), 9–20. (§10.14)

[Alf2]   Manuel Alfaro, Mark Conger, Kenneth Hodges, Adam Levy, Rajiv Kochar, Lisa Kuklinski, Zia Mahmood, and Karen von Haam. The structure of singularities in Φ-minimizing networks in $\mathbf{R}^2$. *Pacific J. Math.* 149 (1991), 201–210. (§10.12, §10.14)

[All]   C. B. Allendoerfer. The Euler number of a Riemannian manifold. *Amer. J. Math.* 62 (1940), 243–248. (§8.5)

[Alm]   F. Almgren. Optimal isoperimetric inequalities. *Indiana U. Math. J.* 35 (1986), 451–547. (§10.5)

[Ben]   Itai Benjamini and Jianguo Cao. A new isoperimetric theorem for surfaces of variable curvature. *Duke Math. J.* 85 (1996), 359–396. (§9.7)

[Ber]   Marcel Berger. *Geometry* II. New York: Springer-Verlag, 1977. (§10.5)

[Bro]   John E. Brothers and Frank Morgan. The isoperimetric theorem for general integrands. *Michigan Math. J.* 41 (1994), 419–431. (§10.6)

[Bru]   H. Brunn. Über Ovale und Eiflächen. Inaugural dissertation, München, 1887. (§10.5)

[Cha]   Isaac Chavel. *Riemannian Geometry: A Modern Introduction*. New York: Cambridge University Press, 1993. (Preface)

[Che1] Jeff Cheeger and David G. Ebin. *Comparison Theorems in Riemannian Geometry*. Amsterdam: North-Holland, 1975. (§6.4, §9.0, §9.3)

[Che2] S.-S. Chern. A simple intrinsic proof of the Gauss-Bonnet formula for closed Riemannian manifolds. *Ann. Math.* 45 (1944), 747–752. (§8.5)

[Cie1] Dietmar Cieslik. Hadwiger numbers in network design. In Proceedings of Network Design: Connectivity and Facilities Location Conference, 1997, edited by D. Z. Du and P. M. Pardalos. *Proc. DIMACS Series in Discrete Math. and Comp. Sci.*, 1997. (§10.15)

[Cie2] Dietmar Cieslik. *Steiner-Minimal-Trees*. Norwood, MA: Kluwer Academic Publishers, 1998. (§10.15)

[Coc]   E. J. Cockayne. On the Steiner problem. *Canadian Math. Bull.* 10 (1967), 431–450. (§10.14)

[Con]   Mark A. Conger. Energy-minimizing networks in $\mathbf{R}^n$. Undergraduate thesis, Williams College, Williamstown, MA, 1989, expanded 1989. (§10.15)

[Cos]   C. Costa. Imersoes minimas completas em $\mathbf{R}^3$ de genero um e curvatura total finita. Doctoral thesis, IMPA, Rio de Janeiro, Brazil, 1982. (See also: Example of a complete minimal immersion in $\mathbf{R}^3$ of genus one and three embedded ends. *Bull. Math. Soc. Bras. Mat.* 15 (1984), 47–54.) (Figure 3.5)

[Cou]   R. Courant and H. Robbins. *What Is Mathematics?* New York: Oxford Univ. Press, 1941. (§10.11)

[Ein]   A. Einstein. Die Grundlage der allgemeinen Relativitätstheorie. *Ann. der Physik* 49 (1916). English translation: The foundation of the general theory of relativity, translated by W. Perrett and G. B. Jeffery. In *The Principle of Relativity*, by H. A. Lorentz, A. Einstein, H. Minkowski, and H. Weyl, 109–164. New York: Dover, 1952. (§7.6)

[Fed]   Herbert Federer. *Geometric Measure Theory*. New York: Springer, 1969. (§10.5)

[Fen]   W. Fenchel. On total curvature of Riemannian manifolds: I. *J. London Math. Soc.* 15 (1940), 15–22. (§8.5)

[Fer]   P. Fermat. *Oeuvres de Fermat*. Paris: Gauthier-Villars, 1891. (§10.11)

[Foi]   Joel Foisy, Manuel Alfaro, Jeffrey Brock, Nickelous Hodges, and Jason
         Zimba. The standard double soap bubble in $\mathbf{R}^2$ uniquely minimizes
         perimeter. *Pacific J. Math.* 159 (1993), 47–59. (§10.7)

[Fre]   Christopher French, Kristen Albrethsen, Charene Arthur, Heather Cur-
         nutt, Scott Greenleaf, and Christopher Kollett (1993 Small Geom-
         etry Group). The planar double Wulff crystal. Williams College,
         Williamstown, MA, 1993. (§10.7)

[Für]   Z. Füredi, J. C. Lagarias, and F. Morgan. Singularities of minimal sur-
         faces and networks and related extremal problems in Minkowski space.
         Proc. DIMACS Series in Discrete Math. and Comp. Sci. 6 (1991), 95–
         106. (§10.15)

[Got]   David Henry Gottlief. Note on Gauss-Bonnet. *Amer. Math. Monthly* 104
         (Jan. 1997), 35. (§8.5)

[Gro]   M. Gromov. Isoperimetric inequalities in Riemannian manifolds. In
         Asymptotic Theory of Finite Dimensional Normed Spaces (Appendix I)
         by Vitali D. Milman and Gideon Schechtman, Lecture Notes in Mathe-
         matics, No. 1200, 114–129. New York: Springer-Verlag, 1986. (§10.5)

[Han]   M. Hanan. On Steiner's problem with rectilinear distance. *J. SIAM Appl.
         Math.* 14 (1966), 255–265. (§10.14)

[Har1]   Philip Hartman. On exterior derivatives and solutions of ordinary dif-
         ferential equations. *Trans. Amer. Math. Soc.* 91 (1959), 277–293. (§6.4)

[Har2]   Philip Hartman. *Ordinary Differential Equations.* Boston: Birkäuser,
         1982. (§6.4)

[Has1]   Joel Hass, Michael Hutchings, and Roger Schafly. The double bubble
         conjecture. *Elec. Res. Ann. Amer. Math. Soc.* 1 (1995), 98–102. (§10.7)

[Has2]   Joel Hass and Frank Morgan. Geodesics and soap bubbles in surfaces.
         Math. Z. 223 (1996), 185–196. (§9.7)

[Has3]   Joel Hass and Roger Schafly. Bubbles and double bubbles. *Amer. Sci-
         entist* (Sept./Oct. 1996), 462–467. (§10.7)

[Has4]   Joel Hass and Roger Schafly. Double bubbles minimize. Preprint (1995).
         (§10.7)

[Hel]   Sigurdur Helgason. *Differential Geometry, Lie Groups, and Symmetric
         Spaces.* Boston: Academic Press, 1978. (§6.4, §9.2)

[Hic]   Noel J. Hicks. *Notes on Differential Geometry.* Princeton: Van Nostrand, 1965 (out of print). (Preface)

[Hil]   Stefan Hildebrandt and Anthony Tromba. *The Parsimonius Universe.* New York: Copernicus, Springer-Verlag, 1996. (§10.1)

[Hof1]   David Hoffman. The computer-aided discovery of new embedded minimal surfaces. *Math. Intelligencer* 9 (1987), 8–21. (Figure 3.5)

[Hof2]   D. Hoffman and W. H. Meeks, III. A complete embedded minimal surface in $\mathbf{R}^3$ with genus one and three ends. *J. Diff. Geom.* 21 (1985), 109–127. (Figure 3.5)

[Hof3]   David Hoffman, Fusheng Wei, and Hermann Karcher. Adding handles to the helicoid. *Bull. Amer. Math. Soc.* 29 (1993), 77–84. (Figure 3.6)

[Hop]   Heinz Hopf. Über die *curvatura integra* geschlossener Hyperflächen. *Math. Ann.* 905 (1925), 340–376. (§8.4)

[How]   Hugh Howards, Michael Hutchings, and Frank Morgan. The isoperimetric problem on surfaces. Preprint (1997). (§9.7)

[Jar]   V. Jarnik and O. Kossler. O minimalnich grafesh obsahujicich n danych bodu. *Casopis Pest. Mat. Fys.* 63 (1934), 223–235. (§10.11)

[Jef]   G. B. Jeffery. *Relativity for Physics Students.* New York: Dutton, 1924. (§7.1)

[Kno]   Herbert Knothe. Contributions to the theory of convex bodies. *Michigan Math. J.* 4 (1957), 39–52. (§10.5)

[Lau]   Detlef Laugwitz. *Differential and Riemannian Geometry.* Boston: Academic Press, 1965. (Preface)

[Law]   Gary Lawlor and Frank Morgan. Paired calibrations applied to soap films, immiscible fluids, and surfaces or networks minimizing other norms. *Pacific J. Math.* 166 (1994), 55–63. (§10.13)

[Lev1]   Mark Levi. A "bicycle wheel" proof of the Gauss-Bonnet theorem. *Expo. Math.* 12 (1994), 145–164. (§8.0)

[Lev2]   Adam Levy. Energy-minimizing networks meet only in threes. *J. Und. Math.* 22 (1990), 53–59. (§10.12)

[Loh]   J. A. Lohne. Essays on Thomas Harriot. *Arch. Hist. Ex. Sci.* 20 (1979), 189–312. (§8.7)

[Mil]   John Milnor. Curvatures of left invariant metrics on Lie groups. *Adv. Math.* 21 (1976), 293–329. (§9.2)

[Mor1] Frank Morgan. Area-minimizing surfaces, faces of Grassmannians, and calibrations. *Amer. Math. Monthly* 95 (1988), 813–822. (§5.2)

[Mor2] Frank Morgan. Calculus, planets, and general relativity. *SIAM Rev.* 34 (Jun. 1992), 295–299. (§7.1)

[Mor3] Frank Morgan. The double soap bubble conjecture. MAA Focus (Dec. 1995), 6–7. (§10.7)

[Mor4] Frank Morgan. *Geometric Measure Theory: A Beginner's Guide, Second Edition.* Boston: Academic Press, 1995. (Preface, Figure 3.3, §5.2)

[Mor5] Frank Morgan. Mathematicians, including undergraduates, look at soap bubbles. *Amer. Math. Monthly* 101 (1994), 343–351. (§10.7)

[Mor6] Frank Morgan. Minimal surfaces, crystals, shortest networks, and undergraduate research. *Math. Intelligencer* 14 (Summer 1992), 37–44. (§10.11)

[Mor7] Frank Morgan. Soap bubble clusters, shortest networks, and minimal surfaces. Amer. Math. Soc. video (1991). (§10.11)

[Mor8] Frank Morgan, Christopher French, and Scott Greenleaf. Wulff clusters in $\mathbf{R}^2$. *J. Geom. Anal.*, to appear. (§10.7)

[Nas]   J. Nash. The embedding problem for Riemannian manifolds. *Ann. Math.* (2) 63 (1956), 20–63. (§6.0)

[Nit]   Johannes C. C. Nitsche. *Vorlesungen über Minimalflächen.* New York: Springer-Verlag, 1975. Partial translation: *Lectures on Minimal Surfaces I.* New York: Cambridge University Press, 1989. (Preface)

[ONe]   Barrett O'Neill. *Semi-Riemannian Geometry.* New York: Academic Press, 1983. (§7.2)

[Pan]   P. Pansu. *Sur la régularité du profil isopérimétrique des surfaces riemanniennes compactes.* Preprint (1997). (§9.7)

[Pet]   Ivars Peterson. Toil and trouble over double bubbles. *Sci. News* 148 (Aug. 12, 1995), 101. (§10.7)

[Sch]   Richard Schoen. Conformal deformation of a Riemannian metric to constant scalar curvature. *J. Diff. Geom.* 20 (1984), 479–495. (§6.0)

[Som]   D. M. Y. Sommerville. *An Introduction to the Geometry of n Dimensions.* New York: Dutton & Co., 1929. (§5.2)

[Spa]   Barry Spain. *Tensor Calculus.* New York: Wiley, 1953. (§7.1)

[Spi]   Michael Spivak. *A Comprehensive Introduction to Differential Geometry,* Vols. I–V. Houston: Publish or Perish, 1979. (Preface, §8.5, §9.3)

[Ste1] Jacob Steiner. Aufgaben und Lehrsätze. *Crelle's J.* 13 (1835), 361–365. (§10.11)

[Ste2] Jacob Steiner. *Über den Punkt der kleinsten Entfernung* (1837). Berlin: Gesammelte Werke, 1882. (§10.11)

[Sto]   J. J. Stoker. *Differential Geometry.* New York: Wiley-Interscience, 1969. (Preface, Exercise 6.5)

[Swa]   K. J. Swanepoel. Vertex of Steiner minimal trees in l_p^d and other smooth Minkowski spaces. *Dis. Comp. Geom.*, to appear. (§10.15)

[Tay1] Jean E. Taylor. Crystalline variational problems. *Bull. Amer. Math. Soc.* 84 (1978), 568–588. (§10.5)

[Tay2] Jean E. Taylor. Unique structure of solutions to a class of nonelliptic variational problems. *Proc. Symp. Pure Math.* 27 (1975), 419–427. (§10.6)

[Wec]   Brian Wecht, Megan Barber, and Jennifer Tice (1995 Small Geometry Group). Double salt crystals. Williams College, Williamstown, MA (1997). (§10.7)

[Wei]   Steven Weinberg. *Gravitation and Cosmology.* New York: Wiley, 1972. (§7.1)

[Won]   Yung-Chow Wong. Differential geometry of Grassmann manifolds. *Proc. Nat. Acad. Sci. USA* 57 (1967), 589–594. (§5.2)

[Wul]   G. Wulff. Zur Frage der Geschwindigkeit des Wachstums und der Auflösung der Krystallflächen. *Zeitschrift für Krystallographie und Mineralogie* 34 (1901), 449–530. (§10.5)

Symbol Index

Name Index

Subject Index